BAY BRIDGE

BAY BRIDGE

HISTORY AND DESIGN OF A NEW ICON

Donald MacDonald and Ira Nadel

CHRONICLE BOOKS

SAN FRANCISCO

Library of Congress Cataloging-in-Publication Data available.

ISBN: 978-1-4521-1326-5

MIX
Paper from
responsible sources
FSC® C104723

Manufactured in China.

Design by Lydia Ortiz

10 9 8 7 6 5 4 3 2 1

Chronicle Books LLC
680 Second Street
San Francisco, CA 94107

www.chroniclebooks.com

CONTENTS

THE TITAN
OF BRIDGES

The January 1933 issue of *Popular Mechanics* featured a new engineering marvel on the West Coast: an innovative bridge that would become the largest suspension bridge in the world upon its completion. With the sobriquet "The Titan of Bridges," the new San Francisco–Oakland Bay Bridge, shortened to the Bay Bridge, would be an engineering and transportation wonder solving the problem of increased population and automobile traffic between the East Bay and San Francisco. Through rapid planning—it took only two years to design and three to build—the new link between Oakland and San Francisco would quickly prove to be a vital element in the economic growth of the area.

Longer than its competitor, the Golden Gate Bridge, which would be inaugurated six months after the Bay Bridge opened in November 1936, the "Titan" remained continuously in use from 1936 until 1989, when the Loma Prieta earthquake caused a fifty-foot roadway section to collapse, closing the bridge for one month and causing widespread concern over its safety.

Surprisingly, this $77 million public works project, undertaken during the Depression, was approved, designed, and built both without a popular vote and with great speed. Indeed, political and social consensus supporting the bridge was a unique feature of an exceptional structure, actually two bridges: the West Span, a double suspension bridge linked by a giant anchorage between San Francisco and Yerba Buena Island, and the East Span, a cantilevered truss bridge between Yerba Buena Island and Oakland. But the purposeful construction of the 1936 bridge is in sharp contrast to the delays, wrangling, and cost overruns of the new East Span. Called the White Span because of its distinctive 525 ft. white tower and roadway, the new bridge did not begin construction until 2004 and will not be completed until September 2013, twenty-four years after the 1989 earthquake.

How all this this came about and how the world's longest single-tower self-anchored suspension bridge became the choice for the new East Span is the story of this book, co-authored and illustrated by Donald MacDonald, the architect who designed its signature structure.

1
WHAT'S
IN A NAME?

Yerba Buena Island is a natural outcropping in San Francisco Bay, approximately one and a quarter miles from San Francisco, and is the touchdown for the East and West spans of the original Bay Bridge. First known as Sea Bird Island, it became Goat Island about 1836 when a Captain Gorham Nye placed goats imported from the Sandwich Islands on the small landmass for sale to trading vessels. He moved them to the island because they had been destroying the flowers in his San Francisco home, but they reproduced so rapidly—by 1849 there were nearly a thousand—the name Goat Island seemed appropriate. Wood Island was the next designation, and sometime later it became Yerba Buena, the name originating from a nearby town because of an aromatic trailing vine that covered the slopes of the town and the island. The plant, whose common name is an English form of the Spanish *hierba buena*, meaning "good herb," is related to the mint family.

The first legislature of California officially established Yerba Buena as the island's name in February 1850—but it didn't stick. By 1895, the U.S. Geographic Board changed it back to Goat Island, which lasted until June 1931. The Geographic Board then reversed itself, restoring the Spanish "Yerba Buena" in response to demands to return to its official, legislatively approved name.

Centuries before the settlement of the Bay Area, Ohlone Native Americans fished in the area and explored the island, traveling there in either canoes or tule barges *(Figures 1 & 2)*. They established a fishing station, as well as a kind of Turkish bath called a temascal in Spanish, considered a remedy for many ailments. Birds, as well as fish, provided food. Once Europeans started to arrive, the island also became a gathering place for fisherman and sailors because of its position in the middle of the uncharted bay; it offered limited shelter and resources, with its highest point some 344 feet above the low-water mark. When William Bernard from the ship *Edward Everett* arrived on the island in 1849, he found nothing but a few goats, although the remains of an extensive Native American village appeared on the east shore *(Figure 3)*. The buildings ranged from the ruins of old houses to bones, shells, and cremating pits. Bernard later returned to live on the island for a short time.

PARTIAL MAP OF SAN FRANCISCO BAY SHOWING
LANDS OF THE OHLONE INDIANS

TYPICAL OHLONE INDIAN REED BOAT USED FOR
2000 YEARS ON SAN FRANCISCO BAY

MARSH LAND

SMOKE HOLE

REED HUT

OHLONE INDIANS SUMMER HOUSING
ON YERBA BUENA ISLAND

The island has had its share of mystery, notably a legend of buried treasure, as smuggling was a particularly successful business in the 1830s, with goods taken from ships arriving in the harbor and hidden on the rocky outcroppings; the alleged items included opium and silver. In 1837, a Spanish sloop returning to Spain with the wealth of the Mission Dolores crashed on the island in a storm. It was likely wrecked on the northern point, with most, but not all, of its treasure saved. Survivors apparently buried much of the church plate and silver on the island, although they never returned to claim it. Coupled with stories from the California Gold Rush of 1849, numerous prospecting parties headed for the island, but rarely with any luck. During a fire in San Francisco, however, a great deal of stolen gold and silver was buried on the island. Suspected to be part of the Mission Dolores treasure, it was later recovered with the help of the police.

Spanish explorers sailed past the inland island but some squatters during the early California Gold Rush called it home. Sailors made excursions to the island from their ships anchored in the bay. By 1852, the government realized the island's strategic importance, proposing to place gun batteries there, thereby including it in a third line of fortifications, joining earlier forts on Alcatraz and Fort Point.

The military reservation of the late nineteenth century became a Quartermaster depot, eventually being transferred to the U.S. Army Corps of Engineers. Officers stationed there later brought their families. The presence of the military personnel and a small community meant the need for a cemetery on the west end of the island, where markers reach back to 1852. Of course, death had always been a presence on the island. Almost fifty years later, a pair of skeletons were found buried in sitting positions, while on the eastern part of the island specimens of stone mortar, stone pipes, and a dog skeleton were located—further evidence that the island had been inhabited for a very long time.

Earlier in the nineteenth century, the death of Captain Edward Lindsey aboard his ship, the *Palmyra*, while it was docked in the bay in August 1852 underscored the importance of the island. An elaborate funeral cortege rowed out from the ship, led by a longboat with four oarsmen, followed by a line of ships' boats headed for the island's graveyard for a full naval burial. The most unusual burial, however, was that of a despondent Italian nobleman who, in the 1850s, after Garibaldi's seizing power from the king of Italy, moved to San Francisco. But even in his new life, he remained discouraged and unhappy and ultimately decided to take his own life. He crossed to Yerba Buena Island one night and dug his own grave, arranging the dirt by means of boards set upon a trigger so that when he fired his gun and fell into the grave, the dirt would tumble on top of him. In this way, he buried himself. When his body was discovered a few days later, however, the coroner insisted it be removed and taken to the city for burial in a pauper's grave.

Also about this time, 1853, one Thomas Dowling sought to establish a home on the island, anticipating a fortune from its strategic position and valuable stone. He began to develop a quarry and built a house and dock. Scows transported the rock from the island to the city.

It was at this point that a private developer recognized the potential in the island and proposed a dry dock for ship repair. The U.S. government denied the request in September 1865, claiming that the navy would seldom require it; the navy yard at Mare Island adequately serviced their ships. A year later, however, the military command at Alcatraz Island decided to send over a small detachment of twelve men to take up a post on the island, by then identified as Yerba Buena. By 1868, the federal government ordered a detachment of 125 men from the Engineering Corps to occupy the island, transported from Point San Jose. Soon, the Central Pacific Railroad and supporters sought to use the island as a railroad terminus—the first of numerous efforts to incorporate

COAST GUARD STATION AND THE BUOY TENDER ON THE
EAST SIDE OF YERBA BUENA ISLAND AS IT IS TODAY

the island into the transportation system of the region—but the military was not open to the island being used in that manner.

The establishment of a lighthouse, considered one of the most important on the West Coast and physically in line with Fort Point and Point Bonita, was completed in 1875. Mariners entering the Golden Gate used it to establish their bearings. (Modernized and automated in 1958, it continues to play a vital role in bay shipping and has expanded to become a lighthouse depot and buoy wharf supplying all lights and buoys for the West Coast from Oregon to Mexico [Figure 4].)

In 1898, Congress appropriated funds for a Naval Training Base on the island. President McKinley approved the allocation and the apportionment of land. Responsibility shifted from the army to the navy but until barracks were built, trainees lived on the USS *Pensacola* anchored alongside the island. Some 175 cadets between fifteen and seventeen years old participated. Within ten years, however, the facility was deemed too limited, and the fleet recommended another location: San Diego, where a new facility was readied in 1923. During World War I, however, there were at one time thirteen thousand men trained on the island.

With the removal of the training facility, the navy's activities shifted to a receiving ship, the USS *Boston*, one of Admiral George Dewey's fleet at the Battle of Manila Bay from the Spanish-American War. Traditionally, a receiving ship is a ship used to receive or house newly recruited sailors before they are assigned to a crew. Moored to the shore dock on the east side of the island, this became the official receiving ship of the West Coast, although most of the business for the navy occurred in an administration building on shore. The island was subsequently shared by three branches of government, with almost 30 acres devoted to the Lighthouse Service, 11 acres to the U.S. Army Coast Artillery Corps as a mine depot, and the remaining 107 acres assigned to the navy as the receiving ship area.

Attempts to beautify the island began in 1886 with the California poet Joaquin Miller encouraging others to plant forest trees on Yerba Buena in double rows in the shape of a Greek cross. In 1904, in celebration of California's first Arbor Day, November 27, 1886, some six thousand new trees were planted on the island.

2 ON THE WRONG SIDE OF THE BAY

Joshua Abraham Norton (1819–1880), self-proclaimed "Emperor of These United States" and subsequently "Protector of Mexico," was born in England. He spent his early life in South Africa, emigrating to California in 1849 to try his luck at the Gold Rush. After an initial business success, he went bankrupt and then left the city but soon returned to declare himself emperor in 1859, issuing his own curency. Mark Twain and Robert Louis Stevenson wrote about him, Twain as the king in *Huckleberry Finn* and Stevenson in his 1892 novel *The Wrecker*.

Among his orders was one to dissolve the U.S. Congress and another declaring that a tunnel and bridge be built to connect San Francisco with Oakland. As late as December 2004, there was a campaign to name the new East Span of the Bay Bridge after the charismatic Norton. The San Francisco Board of Supervisors approved the motion eight to two; the Oakland City Council did not.

Norton's prescient (but slightly comic) order to build a tunnel and bridge had a ring of visionary truth to it. The proclamation reads in part

WHEREAS, we issued our decree ordering the citizens of San Francisco and Oakland to appropriate funds for the survey of a suspension bridge from Oakland Point via Goat Island; also for a tunnel; and to ascertain which is the best project; and whereas the said citizens have hitherto neglected to notice our said decree . . . [We] now therefore, do hereby command the arrest by the army of both the Boards of City Fathers if they persist in neglecting our decrees.

Given under our royal hand and seal at San Francisco, this 17th day of September, 1872.

On the western side, the proposed bridge was to go from Yerba Buena to Telegraph Hill in San Francisco; on the east, from the island to the Oakland docks. Few attended to his vision, at the time thought to be wildly impractical. (And yet, Norton was renowned, quite rationally, for imposing a fine of $25 on anyone if they uttered the word "Frisco.")

All this was not a joke. That same year, a Bay Bridge Committee examined the realistic possibilities of bridging the divide, partly driven by the realization that San Francisco was on the wrong side of the bay—at least as far as the railway

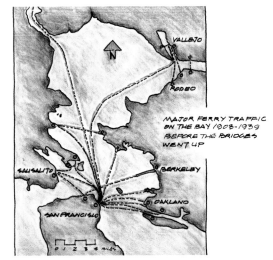

was concerned. In 1869, the transcontinental railroad had been completed, but it ended in Oakland. Federal officials from the U.S. Coast Survey and the Army Engineer Corps reported to the committee that a bridge would be technically feasible but cost would vary if it were a railway or railway-and-highway bridge. A solid causeway between Yerba Buena Island and Oakland, however, would cause "irreparable injury" to the harbor, interfering with the bay's tidal current.

Proposals had begun as early as 1851 when William Walker, editor of the *San Francisco Herald*, proposed the idea of a bridge. In 1853, an impresario in Gold Rush California,

Tom Maguire, held a pageant that showed a massive suspension bridge across the bay as a backdrop. Three years later, in 1856, a comic poem by Lieutenant George H. Derby referred to "The Song of the Oakland Bridge," the same year the California legislature introduced a bill calling for such a bridge; it died for lack of interest. But even if the transcontinental railway ended in Oakland, the fear that if there was not a bridge to San Francisco the city would no longer be the regional trade center ran high.

An anonymous pamphlet entitled *The Railroad System of California* (1871), possibly written at the request of the Central Pacific Railway, outlined plans for the construction of a causeway linking Oakland and Yerba Buena Island. In April 1872, a real estate circular noted that the Bridge Committee had submitted its report to the Board of Supervisors, recommending a compromise with the Central Pacific Railway. Bridging the bay would occur, perhaps, at Ravenswood (the touchdown for the Dunbarton bridge today).

BERTHS   PIER   BERKELEY   PIEDMONT   OAKLAND

RAIL KEY SYSTEM SYMBOL
ORIGINATED IN 1903

At that time, an elaborate and successful ferry system controlled bay transportation. Starting as early as 1850, it offered regular and reliable service between Oakland and San Francisco (*Figure 5*). Railways, like the Southern Pacific, Santa Fe, Western Pacific, and Key System, initially provided the ferry service, since there was no direct rail link to San Francisco from Oakland. (In 1914, air ferries—planes—were tried, an effort repeated in 1930–1933 using seaplanes, but the venture was short-lived because of its high cost.)

By 1920, almost seventy-four thousand people a day used the three Southern Pacific ferry routes linking the Oakland Pier, Alameda Pier, and Oakland Harbor. And although there were few mishaps, there were several severe accidents: In February 1928, for example, because of improper water ballast, the *Peralta* plunged into a swell off of Yerba Buena Island, and thirty passengers were swept into the waters; five drowned. In 1933, a fire in the terminal on the Key System pier off of Oakland destroyed the terminal and gutted the *Peralta* and fourteen interurban electric trams. Proponents of a new bridge naturally maximized these disasters.

The world's largest ferry boat at the time, the *Solano*, built for these waters, actually carried entire trains, but carrying passengers became more profitable. The ferries soon controlled the commuter traffic, as well as offering

SOUTHERN PACIFIC FERRY BOAT "OAKLAND"

NORTHWESTERN PACIFIC · SOUTHERN PACIFIC'S
"EUREKA"

Left: *Figure 9*
Opposite: *Figure 10*

SAN FRANCISCO MAIN FERRY TERMINAL

newsstands, restaurants, candy stores, and, when permitted, small bars. At one point, twenty-seven ferry companies operated in the Bay Area, including those of a number of railways, the most notable the California Pacific and the Southern Pacific. The Key System, so called because its streetcar route in the East Bay took the shape of a large key *(Figure 6)*, began ferry service in 1903 and operated two of the largest vessels, the *Oakland* and the *Eureka*, transporting first people and then cars to the Ferry Building in downtown San Francisco *(Figures 7 & 8)*.

Naturally, the ferry companies strongly lobbied against any bridge, but with the increased use of the automobile and limited capability of the ferries, pressure mounted on the government to provide reliable and efficient access back and forth across the bay, especially between San Francisco and Oakland *(Figures 9 & 10)*.

The economic clout of the railways and their control of the ferry system prevented progress on a bridge. Additionally, the bridge remained too much of an engineering and economic challenge because the bay was both too

DOCKED FERRY

KEY ROUTE TERMINAL · OAKLAND ·

wide and too deep. In 1914, for example, the War Department rejected a proposal by a San Francisco engineer, F. E. Fowler, for a cantilevered bridge because it would be a barrier to naval traffic. In 1921, a tunneling project was suggested—but it would likely have been inadequate for automobiles. But the very popularity of the automobile began to exert continued pressure on the government to provide a dependable and speedy transportation scheme to connect the East Bay with San Francisco.

3 TRAFFIC JAMS & BLESSINGS

With a growing population and the development of competing ferry systems throughout the bay causing confusion and disarray, the transportation system was disorganized. The ferries, though large, were soon too small for the increasing number of commuters, and crossings were often delayed because of weather. Mass transit could only deposit its passengers at, say, the Oakland Pier, for transport across the bay. The Key System did little more than shuttle commuters from the outlying communities to the Oakland piers, where they waited for Key System ferries to take them to San Francisco. A bridge, one that would include some form of train service, became increasingly necessary, as preliminary planning showed in 1921 when Ralph Modjeski, at the time one of the best-known bridge engineers in America, led a team to make an initial survey and take minor borings for a bridge. Funds for this work came from the San Francisco Motor-Car Association. Later, several private firms put forward plans for a bridge to the City of San Francisco.

Herbert Hoover, then U.S. secretary of commerce and an engineer as well as a Californian, was instrumental in the development of the project. For him the bridge was a priority, but he had to overcome military objections concerning security and the safe passage of ships. Testimony in Washington by San Francisco's mayor James Rolph in 1928 challenged the military's objections and argued for an Oakland–San Francisco link via Yerba Buena Island, given the anticipated growth of the East Bay. Further testimony from Michael O'Shaughnessy, a San Francisco city engineer who was influential in the establishment of the Golden Gate Bridge, and from a harbor pilot, described growing bay congestion and that the recent drowning of five passengers from the *Peralta* only signaled future disasters.

In 1929, California created its Toll Bridge Authority, making it and, of course, the state, responsible for the construction of a bridge across the bay, at the same time as newly elected President Hoover and California governor, Clement C. Young, established the Hoover-Young Bay Bridge Commission. They undertook a comprehensive technical study of the feasibility of a bridge and in 1930 delivered its report, recommending that a bridge be built.

Hoover had long supported such a project to rejuvenate the economy and transportation system of the East Bay. It would also give people jobs during the Depression. To make the project feasible, a route from Emeryville via Goat (Yerba Buena) Island to Rincon Point in San Francisco was chosen. Such a path would reduce the amount of material and construction labor needed. But the U.S. Navy, which had a base on the island, controlled the land, and congressional approval was necessary for the island to be used. Intense lobbying occurred before California finally received the go-ahead in February 1931; it was now officially a state project authorized by the legislature but aided by the federal government.

A new Bay Bridge Division of the California Department of Public Works, led by Charles H. Purcell, Charles Andrew, and then Glenn B. Woodruff, began to assemble a team of fifty engineers, along with secondary staff such as surveyors and draftsmen. As a check on the design, Purcell also established a Board of Consulting Engineers of distinguished men in their field. This was common practice but may have also protected the initial team, since no one other than Woodruff had any experience in designing or building a bridge of this magnitude. Ralph Modjeski chaired the consultative committee. At one time Joseph Strauss, chief engineer of the Golden Gate Bridge, worked as Modjeski's assistant in his Chicago office.

Challenges were immediate for the 4½-mile length, beginning with construction of deep-water foundations and a long-span suspension bridge. Even alignment with Goat Island, as Yerba Buena Island was then called, posed problems, with four alternatives presented. The biggest issues for the consultative board were (1) the design for the West Span needed to cross the 2-mile divide between San Francisco and Goat Island, (2) how to sink foundations into the bay, and (3) the bridge type possible for the long East Span. Regulatory concerns came from the War Department, which had specific requirements for vertical and horizontal clearances for shipping and military needs.

How to span the distance then, while minimizing the number of piers to reduce the expense of the work, was difficult. And no serious design work could begin until resolution of the West Bay crossing and its foundations.

The largely unprecedented choices for the West Span were either multiple cantilever spans, a double-suspension span with a center anchorage, a continuous suspension bridge with two anchorages but more than two towers, and, finally, a conventional suspension bridge with two anchorages and two towers. (A cantilever design had been the basis of the Hoover-Young Commission's plans, as well as those of the War Department and the Reconstruction Finance Corporation, who would supply funds.)

The move to a suspension bridge for the West Span began, however, almost as soon as the Hoover-Young report—which focused on location, the basic geometry of the structure, and finances—was tabled. But the length of the main suspension span, longer than the George Washington Bridge, posed problems. The distance of the main span—4,100 feet—would almost equal that of the Golden Gate. But the total length of the West Span exceeded that of the Golden Gate, with the side spans extending 2,000 feet each, nearly twice as long as the side or approach spans of the Golden Gate. This was a heroic challenge, partly because no bridge of such length had ever been built. Winds and earthquake risks were further considerations. The final selection of the twin-suspension design over a continuous span occurred for engineering—as well as aesthetic—reasons. A single-suspension span, massive and potentially unstable because of uncertain pier construction, did not seem feasible. It was also $2 million higher in cost.

The entire bridge had to fit a budget of only $77.2 million, limiting creative choices, although at that point it was the most expensive single public works project to date in America. But unlike other great bridges in America noted for individual architects or engineers—John A. Roebling for the Brooklyn Bridge, Ralph Modjeski for the Ben Franklin

Bridge (Philadelphia), and Joseph Strauss for the Golden Gate—the Bay Bridge did not have a single, guiding hand. It almost seems to have been designed anonymously, although Charles H. Purcell, the chief engineer, got credit in the popular press. Later accounts added the name of Glenn Woodruff, the engineer of design, and sometimes Ralph Modjeski. The mix of styles—double suspension for the West Span, cantilevered and then trusses for the East—suggest that the actual designer was a committee. No visible personality was in control.

The engineers who guided the process had three criteria: structural integrity, economy, and aesthetics for a long and complex structure divided by an island in the middle and requiring a massive anchorage midway on its western span. Yet the very lack of cohesiveness contributed to its distinctiveness: The unusual back-to-back suspension span, its use of a variety of bridge types (a 1,400-foot cantilevered span linked to a 291-foot truss span on the East side), and the inability to

anchor the piers of the East Span to bedrock created a disjoined aesthetic.

The design of the bridge was done in a hurry. The Hoover-Young Commission made important locational decisions accelerating the design process, which essentially took twenty-four months. By early 1933, bridge specifications for what would be at the time the longest and most expensive bridge in the world were complete and went out to bid. Astonishingly, within three years, it was built.

The choice of a suspension bridge, on the west side from San Francisco to Yerba Buena Island and then a cantilevered truss system on the east (prefabricated sections extending out via cranes and fitted into place) until the viaduct continuing to Oakland, had as much to do with expediency as with cost. Politics, however, played a crucial role, as Charles H. Purcell, masterminding and maneuvering to assemble a Bay Area coalition, wrangled with the army, the navy, and the Reconstruction Finance

Corporation while battling with the railroads and ferry companies. And yet, miraculously, completion of the bridge occurred *under* budget and *ahead* of schedule, perhaps due to the unanimous support for the bridge from every city and county in the region. But such unanimity would begin to dissolve after 1936, and by 1989, the year of the Loma Prieta earthquake that severely damaged the bridge, rivalries would be a major stumbling block to replacing the aging structure.

Nevertheless, construction in 1933–1936 was problematic. The West Span was a challenge, with the bay up to one hundred feet deep in places, and the soil requiring new foundation-laying techniques. At the time, suspension bridges could not have more than two towers because of stability considerations. A two-tower span 1.8 miles between anchorages was impractical; the solution was to construct a massive concrete anchorage halfway between San Francisco and Yerba Buena, building a complete suspension bridge on either side of the anchorage. As a result,

the West Span has an unusual configuration of twin suspension bridges or four towers (each 519 feet tall), two on either side of the tall, massive anchorage. The total length is 10,304 feet. Height was also a consideration: The mid-channel had to be at least 220 feet above the bay for all ship traffic. Horizontal clearances were initially too short, as the War Department and shipping interests made clear. Greater clearance was needed, which profoundly affected the choice of bridge designs. Height and horizontal clearances determined form.

Pressure on the state to involve architects in the project grew, especially after the chief state engineer, Charles H. Purcell, published a sketch of a long, four-towered suspension bridge followed by a cantilevered truss structure past Yerba Buena Island toward Oakland. The California architect Timothy Pflueger saw an opportunity to obtain some new work with the proposed bridge. The Depression was affecting his firm severely, despite his preeminence:

PACIFIC TELEPHONE AND
TELEGRAPH BUILDING .1925

He had designed San Francisco's first modern skyscraper, the twenty-six-story Pacific Telephone and Telegraph Company headquarters in 1925 *(Figures 11 & 12)*. Pflueger pursued Purcell, and by 1933 a committee of consulting architects was appointed to work with the bridge engineers, with Pflueger as chair.

This Board of Consulting Architects began its work, but the utilitarian nature of the Bay Bridge limited their creative input, unlike the contribution of Irving Morrow, the architect who had assisted on the Golden Gate Bridge. Among other contributions, he added the step-backs of its soaring towers. Morrow, however, was brought in early in the design phase of the Golden Gate Bridge, before engineers had made any decisions. Pflueger and the two other architects who formed the consulting board came in later and had little influence on stylizing the design of the Bay Bridge, especially its towers *(Figure 15)*.

Although there were eight proposals for the West Span and six for the East, budgetary restrictions prevented much design elaboration *(Figures 13 & 14)*. State engineers, in fact, were well ahead of the architects and appear to have controlled the design process.

Disputes, disagreements, and discord defined engineer/architect relations on the Bay Bridge. The extra costs and additional build time were given as the reasons for the rejection of key aesthetic changes. Purcell only had $77 million, and that amount would not change. For example, efforts to add two feet to the height of the tower columns, increase the size of the tunnel portal, and alter a portion of the bracing were rebuffed: The additional $150,000 needed to add two feet of height would itself be costly—and besides, the steel mill already had the specs and order to construct the towers. Engineers did agree, however, to the bolder diagonal bracing without the horizontal struts suggested in earlier drawings *(Figure 15)*.

ARCHITECTURAL DETAILS AT
THE UPPER LEVELS OF THE
PACIFIC TELEPHONE · TELEGRAPH BUILDING

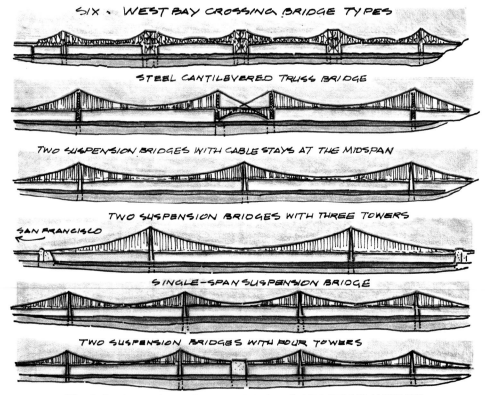

SIX · WEST BAY CROSSING BRIDGE TYPES

STEEL CANTILEVERED TRUSS BRIDGE

TWO SUSPENSION BRIDGES WITH CABLE STAYS AT THE MIDSPAN

TWO SUSPENSION BRIDGES WITH THREE TOWERS

SAN FRANCISCO →

SINGLE-SPAN SUSPENSION BRIDGE

TWO SUSPENSION BRIDGES WITH FOUR TOWERS

TWO SUSPENSION BRIDGES WITH ANCHORAGE HOUSING
AT MIDSPAN AS CONSTRUCTED

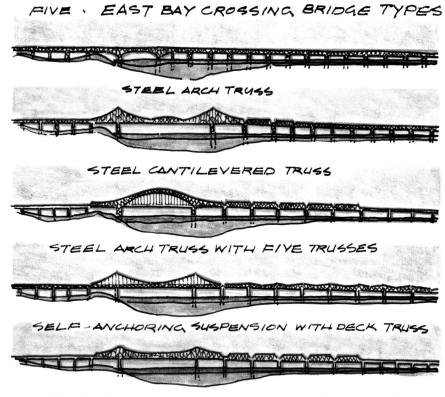

FIVE · EAST BAY CROSSING BRIDGE TYPES

STEEL ARCH TRUSS

STEEL CANTILEVERED TRUSS

STEEL ARCH TRUSS WITH FIVE TRUSSES

SELF-ANCHORING SUSPENSION WITH DECK TRUSS

STEEL CANTILEVERED TRUSS WITH DECK TRUSSES
AS CONSTRUCTED

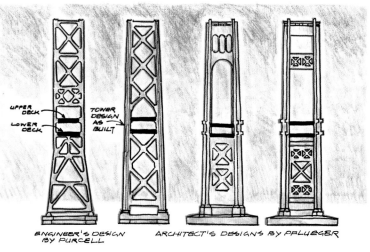

UPPER DECK

LOWER DECK

TOWER DESIGN AS BUILT

ENGINEER'S DESIGN BY PURCELL

ARCHITECT'S DESIGNS BY PFLUEGER

FOUR · 1933 · STUDIES FOR SUSPENSION TOWERS

Similarly rejected were plans for a Roman-style viaduct and alternate versions of concrete arches for on- and off-ramps in San Francisco. An exasperated Pflueger at one point exclaimed, "Isn't there anything to this bridge but cheapness?" The architect's nemesis seemed to be Glenn Woodruff, the state's design engineer.

The battles continued, with the center anchorage next. Pflueger hired the artist/architect Ralph Stackpole to design the 197-foot-above-water structure, which also descended 180 feet below. The idea was to fashion an Egyptian-like figure on the exterior, an art deco–style sculpted object to hold the cables. Of course, it was rejected in favor of a massive, undecorated slab that dominates the surrounding channels (Figure 16).

Sculptures of eagles and panels depicting motion for the San Francisco and Yerba Buena anchorages were similarly refused. Engineers did not want any sculpture on the bridge, as Ralph Modjeski, chairman of the board of consulting engineers, made clear to Purcell. Pflueger was successful, however, with the frontage of the tunnel

to Yerba Buena Island heading west from Oakland. A formidable entrance for the world's largest-diameter tunnel was thought to be appropriate. An arched portal with three stepped concrete archways was accepted (*Figure 18*).

A utilitarian cantilevered truss for the East Span had more to do with foundations than politics or money. The bay was simply too unstable to support large piers because the bottom was mud, not bedrock—and it was prohibitively expensive to bore down to bedrock for foundations, although the performance of piers in mud—actually old-growth fir trees—was not known. The geology of the bay floor required the cantilever-truss system, just as the suspension section was chosen because of the need to maintain a certain height to preserve shipping channels and the ability to locate the massive foundations in bedrock. The bay floor on the west side would also support heavy foundations for long span structures, including the huge center anchor, because of bedrock.

CONCRETE HOUSING WITH CABLE ANCHORAGE INSIDE

OAKLAND CRANES

CENTRAL ANCHORAGE BLOCK FOR THE TWO WEST SUSPENSION BRIDGES

Proposed for the East Bay was a 1,400-foot cantilever just off the shore of Yerba Buena Island, followed by four 500-foot truss spans and then a long stretch of fixed spans of shorter length to the Oakland shore. The unstable abutments and need to reduce the weight of the structure necessitated the cantilever-truss structure. The War Department specified the approximate lengths of these spans. A tied-arch form was also considered but rejected

FIRTH OF FORTH BRIDGE · SCOTLAND · 1890
'CANTILEVERED TRUSS'

because of cost, some $600,000 more than a cantilever-truss design. Concerns over economy and structural integrity favored the cantilever, one of the best-understood bridge designs at the time. But the absence of a coherent, sustained form, caused by challenges of length and footings, meant that each element performed a specific solution to a specific problem and mitigated any effort at spatial harmony.

The design team realized that their mixed design was not as aesthetically pleasing as other forms, although the great Firth of Forth Bridge was a distinguished example of a cantilevered truss (Figure 17). One other early design,

considered but turned down by state engineers, was a self-anchored suspension bridge. Its appearance and expense led to its rejection, although this would become the very design—refined and incorporating new technologies—chosen in 1998 for the new East Span. But despite being limited by money and a shortened design time to build the longest bridge, with the deepest piers, ever attempted, the results were impressive.

One final element of the bridge needed a decision: the color. State engineers favored black, which civic groups opposed. Purcell, however, argued that black would highlight the structural complexity of the bridge and in that way the structural work would be more visible. Further discussion expanded the choice to either black or aluminum. Gray also became an alternative, a color used on the George Washington Bridge (the towers are gray, its cables aluminum). Grudgingly, the Board of Consulting Architects recommended aluminum, although they called it thin and sheet-metal-like in appearance.

ART DECO TUNNEL ARCH DESIGNED BY
ARCHITECT TIMOTHY PFLUEGER 1937

They actually preferred gray. The architects, however, had few of their suggestions accepted and the final choice was a compromise: aluminum gray.

Not only was there a bridge to design, but also a tunnel that would connect the West and East Spans via Yerba Buena Island. The result was the largest bore tunnel in the world. A segmental arch with vertical sidewalls and an arched roof, the tunnel measured seventy-six feet across at its widest and fifty-eight feet at the crest of the arch. The tunnel arch, distinctly art deco, is integrated into the island's hillside *(Figure 18)*.

The two levels with five lanes of traffic in each direction correspond to the two deck levels of the East and West Spans, originally one for cars, the other for trucks and trains *(Figure 19)*.

The process of tunneling involved dynamite blasting, drilling, and mucking (removal of dirt and debris). A temporary rail track aided in the removal of material that became the initial landfill for Treasure Island (named after Robert Louis Stevenson's novel; Stevenson spent 1879–1880 in San Francisco) under the authority of the Army Corps of Engineers. There were not many major incidents in the construction of the tunnel, which could not be said of the bridge, where some twenty-seven men died, some falling into the concrete as it was being poured for the anchorage and foundations. Work could not stop to extricate them. Over the three years it took to complete, nearly eight thousand men worked on the bridge project. At its opening, it became the largest cantilevered truss span

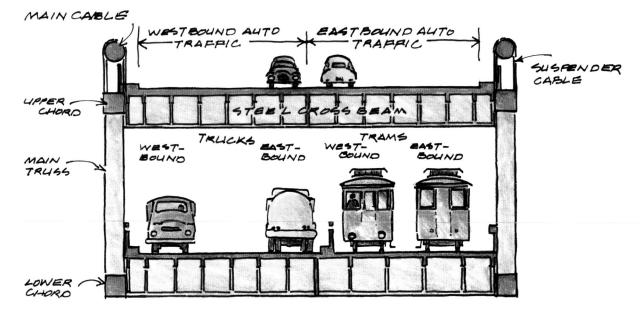

MAIN CABLE

WESTBOUND AUTO TRAFFIC

EASTBOUND AUTO TRAFFIC

SUSPENDER CABLE

UPPER CHORD

STEEL CROSS BEAM

TRUCKS

TRAMS

WEST-BOUND

EAST-BOUND

WEST-BOUND

EAST-BOUND

MAIN TRUSS

LOWER CHORD

SECTION THRU THE 1937 BAY BRIDGE
SHOWING DIVIDED TRAFFIC LANES

TYPICAL TRAM OPERATING ON THE RAIL KEY SYSTEM
FROM OAKLAND TO SAN FRANCISCO

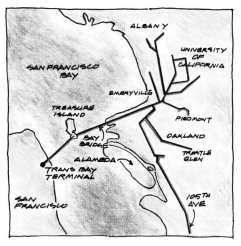

RAIL KEY SYSTEM OPERATED
TO SAN FRANCISCO 1937-1959

and the longest double-suspension bridge in the world. "White Magic" was the term favored by the press for the impressive structure "flung across our waters," as the *San Francisco Recorder* declared. In short, it was a phenomenon.

The opening of the bridge in November 1936 was a gala event for the entire Bay Area. Preceding the opening celebrations was an unexpected papal blessing, performed by Eugenio Cardinal Pacelli, Vatican secretary of state. On a two-day stop in San Francisco several weeks before the official opening, he insisted on seeing the largely completed structure. Accompanied by Archbishop J. J. Mitty of San Francisco, he stopped at the center pier of the West Span and blessed the bridge. (Within three years, Cardinal Pacelli would become Pope Pius XII.)

Public celebrations occurred between November 11 and 15 of 1936, November 12 being the day the bridge actually opened to traffic. Opening day began with the completion of a boat race from Sacramento and Stockton, which terminated at Pier 3 in San Francisco. At 10 A.M., Five hundred pleasure and working boats left San Francisco to form a "Marine Parade" to Yerba Buena and back, while 250 aircraft from navy carriers in the bay began a massed flight from San Mateo to San Rafael. The afternoon saw air shows, a yacht regatta, an air parade of *China Clippers* at 3 P.M., and navy ship races.

The official lighting of the bridge occurred at 5:30 P.M., followed by celebrations throughout the evening, including a navy ball at the Fairmont Hotel and a public ball at the Oakland Auditorium. November 13 saw a huge parade down Market Street between the Ferry Building and City Hall, with marching bands and drill teams from all over the United States. On the evening of November 14, there was another parade in San Francisco with a "history of bridges" theme, which included man's first bridge, as well as the Pont Neuf, Horatio at the bridge, and similar subjects. The large attendance at these events contributed to the first problem encountered on the bridge: congestion.

Huge traffic jams occurred because people were so eager to drive over the bridge and witness the events. It was estimated that 150,000 cars crossed the bridge in the first thirty-six hours, paying a toll of 65¢ each way. "A city gone mad!" read one report in the *San Francisco Chronicle* regarding the parade on the 13th. Enthusiasm for the transbay bridge was extraordinary. The cover of the colorful celebration day program highlighted a rainbow arching behind a looming suspension tower, with the Ferry Building in the background and the roadway bisecting the image *(Figure 23)*.

TRANS BAY TERMINAL · DEMOLISHED IN 2011.

The faces of hundreds of people formed the bottom margin above the November dates. Hyperbole went unchecked, former President Hoover declaring that the Bay Bridge was "the greatest bridge yet erected by the human race."

The original bridge lacked a rail line, although by 1939 the Key System, with its electric trams, began service on the lower deck *(Figures 20, 21 & 22)*. This system of trams formed the backbone of mass transit for the East Bay. Its

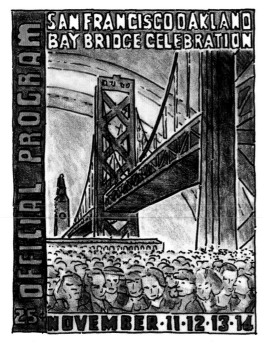

OPENING DAY PROGRAM COVER
NOVEMBER 12TH 1936

westernmost destination was downtown San Francisco's Transbay Terminal, which meant that the company got a coveted spot on the bridge. However, the addition of trams on the bridge ultimately lead to the Key System's demise.

As car ownership in California boomed, interurban rail had no place and certainly not on the bridge, which needed additional space for automobiles. By 1958, automobile lanes replaced the Key System, and its tracks were torn up.

In 1939, the Golden Gate International Exposition celebrated both the Bay (1936) and the Golden Gate (1937) Bridges. The fair was the third momentous public work on the West Coast in a three-year period, and the event was as much a job-creation project as a celebration of the completion of the two bridges, guaranteed to bring dollars from tourists. The creation and extension of the artificial new four-hundred-acre Treasure Island at this time had the goal of developing a regional airport

CHINA CLIPPER TAKING OFF FOR THE ORIENT
FROM YERBA BUENA ISLAND

after the exposition. It took eighteen months of dredging to form the island, which employed twelve hundred landscapists, gardeners, and workers and included the planting of four thousand trees.

The two centers of the exhibition were the Palace of Fine Arts and the Court of the Pacific. The former, an artifact of the Panama-Pacific International Exposition of 1915 in San Francisco, exhibited works from Europe, the United States, and Latin America. The latter, on the island itself, was dominated by an eighty-foot statue of Pacifica designed by Ralph Stackpole and expressed the San Francisco Bay Area's desire to be at the center of the Pacific Basin. Even the Pan Am base for its famous *China Clipper*, which could carry forty-eight passengers across the Pacific in comfort if not luxury, moved from Alameda to Treasure Island to be part of the excitement *(Figure 24)*. *Clipper* operations actually became a new attraction: Tickets were sold to watch maintenance work from a balcony at the Treasure Island hangar.

Only the dark clouds of war diminished the expectations of the exposition.

4 EARTHQUAKE!

For fifty-three years the Bay Bridge remained opened, each year showing an increase in vehicular traffic and a corresponding increase in anxiety, as earthquakes began to affect the region. The Sylmar quake of February 9, 1971, registering 6.6 on the Richter scale, for example, caused $500 million in damages in the San Fernando Valley and killed sixty-five. In response, Caltrans (California Transportation Authority) established a seismic-retrofit program for bridges. And then, at 5:04 P.M. on October 17, 1989, minutes before the scheduled start of the third game of the World Series in San Francisco between the Oakland A's and the San Francisco Giants, a massive quake rocked the California coast from Monterey to San Francisco.

The 7.1 quake severely shook the San Francisco and Monterey Bay regions. The epicenter was near Loma Prieta peak in the Santa Cruz Mountains, approximately nine miles northeast of Santa Cruz and sixty miles south-southeast of San Francisco. The earthquake occurred when the crustal rocks comprising the Pacific and North American Plates abruptly slipped as much as seven feet along their common boundary, the San Andreas Fault system. The rupture occurred at a depth of eleven miles and extended twenty-two miles along the fault, but it did not break the surface of the Earth.

The Loma Prieta earthquake killed 63 people, injured 3,757, and caused an estimated $6 to $10 billion in property loss. It was the first large temblor to jolt the urban region since the Great San Francisco Earthquake of 1906 (magnitude 7.8), which resulted in massive fires and the near destruction of the city. Although the Loma Prieta earthquake struck on the periphery of the region—more than fifty miles from San Francisco and Oakland—it exposed the vulnerability of the area to enormous losses in future quakes, some of which were predicted to be larger and closer to the urban core of the region.

The San Francisco Bay Area lies on the boundary zone between two of the major tectonic plates that make up the Earth's outer shell. Known as the San Andreas Fault, slightly to the west of San Francisco, and the Hayward Fault, running through the Oakland region, the motion of the plates strains the crustal rocks of the region, storing energy that will eventually be released in earthquakes (*Figure 25*).

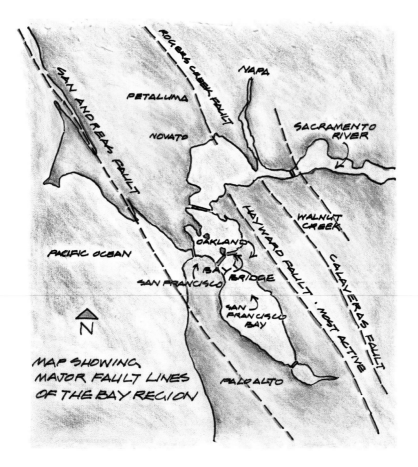

MAP SHOWING
MAJOR FAULT LINES
OF THE BAY REGION

In 1990, the U.S. Geological Survey calculated a 23 percent chance of a 7.5 quake occurring on the San Andreas Fault within the next thirty years. The same calculations predicted a 28 percent chance of a similar magnitude on the Hayward Fault by 2020. On the tenth anniversary of the Loma Prieta earthquake, October 17, 1999, the U.S. Geological Survey released a study predicting a 70 percent chance of a major (6.7+) earthquake striking the Bay Area in the next thirty years.

More than 70 percent of the sixty-three deaths of the Loma Prieta earthquake, including one on the Bay Bridge, occurred in the Oakland–San Francisco region. The bridge death was the result of a seventy-six-by-fifty-foot section of the upper deck on the eastern cantilever side collapsing onto the lower deck. The break occurred at the interface of the cantilever and the truss section. The quake also caused the Oakland side of the bridge to shift seven inches (18 cm) to the east and caused the bolts of one section to shear off, sending the 250-short-ton (230-ton) section of roadbed crashing down like a trapdoor (*Figures 26 & 27*).

When a part of the roadway collapsed, a few upper-deck motorists drove into the hole but landed safely on the lower deck. Traffic on both decks halted, blocked by the section of roadbed. Police began unsnarling the traffic jam, telling drivers to turn their cars around and drive back the way they had come. Eastbound drivers stuck on the lower deck between the collapse and Yerba Buena Island were routed to the upper deck and the westward-bound back to San Francisco. A miscommunication made by emergency workers at Yerba Buena Island, however, routed some of the drivers the wrong way: They were directed to the upper deck, where they drove eastward toward the collapse site. One driver did not see the open gap in time; the car plunged over the edge and smashed onto the collapsed roadbed. The driver died and a passenger was seriously injured.

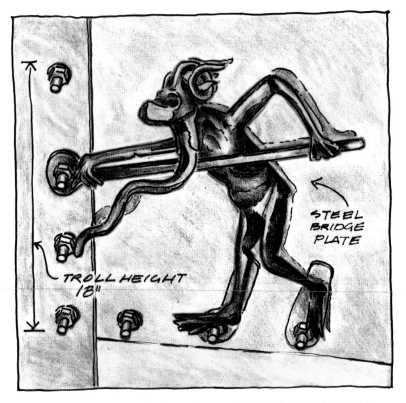

TROLL HEIGHT 18"

STEEL BRIDGE PLATE

BAY BRIDGE TROLL • HIS HOME IS ON THE NORTH SIDE OF THE UPPER DECK OF THE EAST SPAN

With the Bay Bridge closed for a month, commuters turned to the Golden Gate Bridge, resulting in increased traffic, with the highest single-day volume of vehicles in its history—162,414—occurring on October 27, 1989, ten days after the quake. At the time of the event, the Bay Bridge was the nation's most traveled bridge. Within a month, Caltrans had removed and replaced the collapsed section, and reopened the bridge on November 18. When rebuilt, a troll was added to the north side of the upper deck and can be seen in *Figure 26*.

Following the Loma Prieta quake, the fear of another more devastating quake threatened the entire region and Governor Deukmejian created an independent board of inquiry to investigate the damage to the bridge and the collapse of the Cypress Structure on I-880 in Oakland, which had caused forty-two deaths. One explanation for the latter was that the causeway foundation of the highway was on mud. The commission

EAST SPAN OF THE SAN FRANCISCO · OAKLAND BAY BRIDGE SHOWING DAMAGE FROM 1989 EARTHQUAKE

CALTRANS EAST SPAN VIADUCT DESIGN PRIOR TO 1997

called for immediate highway improvements and either the renovation or replacement of the bridge, but politics, money, and aesthetics quickly came into play to delay the process.

The seismic retrofit of the state's bridges and highways became an urgent requirement of the new governor, Pete Wilson. The Northridge earthquake of January 17, 1994, a 6.7-magnitude event in the Los Angeles area causing fifty-seven deaths, accelerated the urgency of the seismic retrofit. Although freeways and their overpasses were the most severely damaged in the Northridge quake, concern for the Bay Bridge intensified, and calls for a speedier process were made. But an October 1994 report indicated that three-quarters of Caltrans's 12,176 bridges still required retrofitting.

By the summer of 1995, a special Seismic Advisory Board suggested that Caltrans replace rather than retrofit the Bay Bridge, the first major study to make this recommendation. A retrofit would be costly, and a replacement bridge, theoretically, not much more. They also authorized a 30 percent design study for a replacement bridge, that is, a study that carries the design work to 30 percent of completion. A 5.4 quake in San Juan Bautista on August 13, 1998, again caused concern. And by now, the Bay Bridge handled some 286,000 cars a day.

Caltrans made their announcement to replace rather than retrofit after a sixty-day period of public review and consultation with the Federal Highway Administration and consideration of five alternatives. Caltrans had to make a final design choice before it could move forward with a final draft of the environmental impact report, but politics quickly interfered. East Bay mayors from Oakland, Emeryville, Albany, Alameda, Piedmont, and Berkeley wanted a more elegant design, claiming that the 7,800-foot skyway viaduct was unattractive *(Figure 28)*. Caltrans listened but cited cost and efficiency, adding that the most sensible choice was a northern alignment. After Caltrans okayed plans to replace the Bay Bridge north of the current structure, new opposition quickly emerged and from some unexpected sources: San Francisco Mayor Willie Brown and Oakland Mayor Jerry Brown, as well as the U.S. Navy.

5 DESIGNING THE WHITE SPAN

ork on the new proposal began with a typical delay. Opposition over financing, and squabbling between state and Bay Area legislators, led to further protests; the governor stating that extra funding for a more aesthetically pleasing bridge must come from Bay Area residents in the form of higher tolls, not state revenues. By March 1997, the Metropolitan Transportation Commission appointed a Bay Bridge Design Task Force to achieve regional consensus on the design of the East Span replacement. It called forth ideas from the public on bridge concepts, three of them appear on the following pages *(Figures 29, 30 & 31)*. After the public hearings, a Request for Proposals was announced.

The Metropolitan Transportation Commission wanted two parallel roadways instead of continuing with the existing double-deck configuration: Stronger seismic stability, economy of construction, and aesthetics were the reasons cited for the change. It would also eliminate the appearance of driving in a cage, since travelers heading east toward Oakland were in the lower section of the cantilever and truss structure. The new plan called for ten lanes with five on each side of the bridge, with two 10-foot-wide shoulders.

Because of deep layers of mud underlying the eastern part of the bay, 85 percent of the new span had to be a causeway elevated on shallow piers *(Figure 32)*.

The Metropolitan Transportation Commission, with public input, also wanted a distinctive cable-supported main span across the deep-water channel adjacent to Yerba Buena Island to be either a self-anchored suspension span or a cable-stayed span. The Request for Proposals, however, required that the projected tower could not be higher than that of the existing West Span and that it be a cable system. The issues, however, were clear: earthquake safety, bridge operations, cost, environmental impact, and aesthetics.

T. Y. Lin International, with Moffatt & Nichol Engineers, was assigned the job of preparing designs. Heading one subgroup, the one that eventually came up with the self-anchored suspension bridge, was the San Francisco architect Donald MacDonald, designer of the country's longest cable-stayed bridge (Cooper River in Charleston, SC). Assisting him was the New York–based design engineer Herbert Rothman of Weidelinger Associates, who had worked on the Verrazano-Narrows Bridge in

EAST BAY

SUSPENSION
CABLE SYSTEM

COMAN·FEHER PYRAMID
SINGLE-TOWER SUSPENSION
BRIDGE

GARY BLACK'S 'SAIL BRIDGE' SINGLE-TOWER
CABLE-STAY BRIDGE

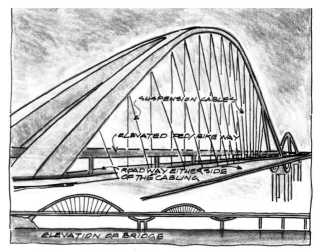

DAVID MORRIS'S ARCHED SUSPENSION
BRIDGE

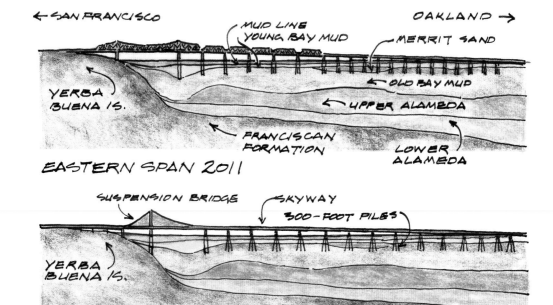

← SAN FRANCISCO

OAKLAND →

MUD LINE
YOUNG BAY MUD

MERRIT SAND

YERBA
BUENA IS.

OLD BAY MUD

UPPER ALAMEDA

FRANCISCAN
FORMATION

LOWER
ALAMEDA

EASTERN SPAN 2011

SUSPENSION BRIDGE

SKYWAY

300-FOOT PILES

YERBA
BUENA IS.

EASTERN SPAN 2013

SOIL PROFILE

VERRAZANO-NARROWS BRIDGE
NEW YORK · BUILT IN 1964

NIMITZ HOUSE ON YERBA BUENA ISLAND

THE TORPEDO SHED ON YERBA BUENA
ISLAND CONSTRUCTED IN 1891

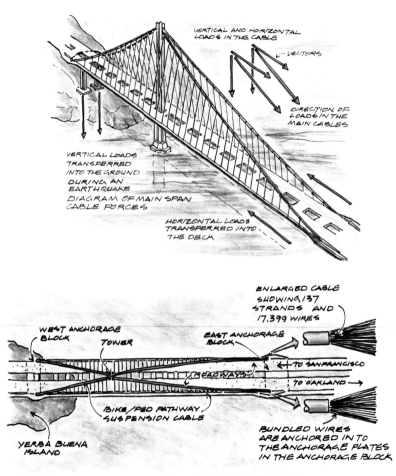

VERTICAL AND HORIZONTAL LOADS IN THE CABLE

← VECTORS

DIRECTION OF LOADS IN THE MAIN CABLES

VERTICAL LOADS TRANSFERRED INTO THE GROUND DURING AN EARTHQUAKE

DIAGRAM OF MAIN SPAN CABLE FORCES

HORIZONTAL LOADS TRANSFERRED INTO THE DECK

ENLARGED CABLE SHOWING 137 STRANDS AND 17,399 WIRES

WEST ANCHORAGE BLOCK

TOWER

EAST ANCHORAGE BLOCK

← TO SAN FRANCISCO

ROADWAYS

TO OAKLAND →

BIKE/PED PATHWAY SUSPENSION CABLE

YERBA BUENA ISLAND

BUNDLED WIRES ARE ANCHORED INTO THE ANCHORAGE PLATES IN THE ANCHORAGE BLOCK

OVERHEAD VIEW OF THE SUSPENSION SYSTEM SHOWING THE ANCHORAGE BLOCKS

New York *(Figure 33)*. The other group, known as H2L2, prepared a cable-stayed design. The same month the RFP was announced, the navy rejected a Caltrans request for site testing, claiming it might damage heritage buildings *(Figures 34 & 35)*, create an unwanted environmental impact, and impede improvements to ramps on Yerba Buena Island.

In June 1997, before any final decision on design, new objections emerged, initially from San Francisco's mayor Willie Brown and the U.S. Navy. Studies showed that a new northern alignment would be necessary for a new touchdown on Yerba Buena Island for the East Span, but Brown opposed the change because he believed it would impede development of the usable land on the island. He also hoped that the adjacent Treasure Island might be developed for condominiums and commercial use. Additionally, he wanted new on- and off-ramps for ease of access for the proposed development. The navy appeared to support him, preventing Caltrans from undertaking soil testing to determine the stability of the land for any proposed new foundations. But preparations for a tower in the likelihood of a suspension bridge could not begin without the soil testing. There was an impasse.

Parties soon divided, the coast guard favoring the northern alignment along with the Design Task Force. By July 1997, even Mayor Willie Brown supported this change, explaining that the economic opportunities of the Port of Oakland outweighed the development of Yerba Buena Island. The Metropolitan Transport Commission also favored the northern alignment, adding that the new bridge be built to a "lifeline" standard, meaning that in the event of a catastrophe, it would be operational for emergency services within three hours. Mayor Brown pledged cooperation, stating that if the navy transferred Yerba Buena and Treasure Island to the city, the city would provide Caltrans with the needed easements for the Bay Bridge. This debate and exchange occurred in September of that year.

By May 1998, a major West Span retrofit began with strengthening the foundations in the water. That same month, however, Oakland voiced its objections to the viaduct portion of the East Bay design, which constituted 85 percent of the span. That same month, the Design Advisory Panel recommended a single-tower, self-anchored suspension bridge. But at that moment, Mayor Willie Brown reversed again, opposing a northern alignment on Yerba Buena Island because it would interfere with the city's development plans for it and Treasure Island.

East Bay leaders then voiced further criticisms that the proposed designs did not establish a gateway to Oakland. They also called for a bicycle/pedestrian lane and light-rail accommodation. Signatories included the mayors of Oakland, Berkeley, Emeryville, Alameda, Piedmont, and Albany. A week later, the Design Task Force chose the self-anchored suspension bridge design, accepting the recommendation of the Engineering and Design Advisory Board, spelling out the critical elements: a low-rise

Oakland approach, a pier-supported viaduct, a signature single-tower, self-anchored suspension span, and a fourth connecting element to the east side of Yerba Buena Island. This would become the world's longest single-tower, self-anchored suspension bridge.

At this time, an Engineering and Design Advisory Panel narrowed their choices to two designs before the final competition took place: a single-tower, self-anchored suspension span (MacDonald Architects' choice) and a single-tower, cable-stayed span (H2L2's choice).

Engineering issues, however, over tension, torsion, ductility, and flexibility emerged in the competition to establish differences. Suspender cables on a self-anchored bridge, for example, attach to the top of the slender four-legged tower and traverse to the outside of the deck *(Figures 36–45)*.

Cable-stayed design has the cables attached to the outside of the deck, with a two-legged tower with an elliptical cross-section that would taper toward the top. Two legs are joined by link beams to provide extra seismic strength, with cables splayed symmetrically from the central tower to the outside of the deck in a semi-fan pattern. The cable arrangement creates a portal through which all traffic passes *(Figures 46 & 47)*.

Detailed arguments pro and con ensued, with numerous and lengthy meetings before the Design Advisory Panel. The process began in November 1997 and ended in June 1998 with the unexpected selection of the self-anchored suspension bridge, designed by MacDonald and Rothman. The vote of the panel, held on May 30, 1998, was twelve to seven recommending the self-anchored form to the Metropolitan Transportation Commission, which endorsed the recommendation.

The panel had been expected to favor the cable-stayed design with cables fanning out from the single tower, but last-minute design changes to the self-anchored form—costing $50 million more than its competitor—persuaded the panel that it was the right design for the space. A more slender tower with two main cables descending to the outside of the parallel roadways created a vaulted, tent-like space through which motorists would pass. And by moving the tower 115 feet closer to Yerba Buena Island, a new dynamic, asymmetrical form dominated the structure to offset the long causeway reaching out to Oakland. A bicycle/pedestrian path was also added on the south side of the span, a foot higher than the roadway, necessary for air flow and stabilization *(Figure 48)*.

MacDonald successfully argued for the self-anchored suspension span partly because the Bay Area had a strong tradition of suspension bridges and the design would complete the necklace of suspension structures around the region *(Figure 49)*. With its distinctive single tower 525 feet high, its convex shape, opposite to the concave form of the main cables, relates the silhouette of the bridge to the shape of the East Bay hills, beyond *(Figure 50)*. The prismatic form of the single steel tower would provide superior earthquake protection, he maintained.

HOODED
SADDLE

NOTE CABLE
ARRANGEMENT

FOUR ROUND CONCRETE TOWER LEGS
WITH AN OPTIONAL CABLE LAYOUT

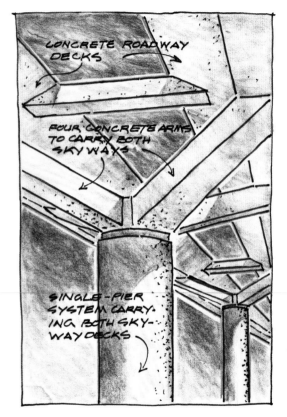

CONCRETE ROADWAY
DECKS

FOUR CONCRETE ARMS
TO CARRY BOTH
SKYWAYS

SINGLE-PIER
SYSTEM CARRY-
ING BOTH SKY-
WAY DECKS

TYPICAL SKYWAY TO MATCH THE
FOUR ROUNDED CONCRETE TOWER
LEGS OF THE MAIN SPAN

Opposite: Figures 38 & 39
Below: Figures 40 & 41

RECTANGULAR PIERS TO MATCH THE DUAL TOWER LEGS SCHEME

TYPICAL SKYWAY VIADUCT TO MATCH THE MAIN SPAN TOWER CONCEPT

DUAL TOWERS WITH RECTANGULAR SHAPED TOWER LEGS SCHEME

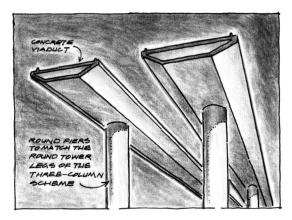

CONCRETE
VIADUCT

ROUND PIERS
TO MATCH THE
ROUND TOWER
LEGS OF THE
THREE-COLUMN
SCHEME

TYPICAL SKYWAY VIADUCT TO MATCH THE
MAIN SPAN TOWER CONCEPT

THREE-COLUMN
SCHEME WITH
VARIABLE DECK
HEIGHTS

MAIN SPAN
TOWER CONCEPT

STEPPED VIADUCT TO MATCH THE THREE-STEPPED TOWER SCHEME

TYPICAL SKYWAY VIADUCT TO MATCH THE
MAIN SPAN TOWER CONCEPT

THREE - STEPPED TOWER LEGS
WITH VARIABLE DECK HEIGHTS

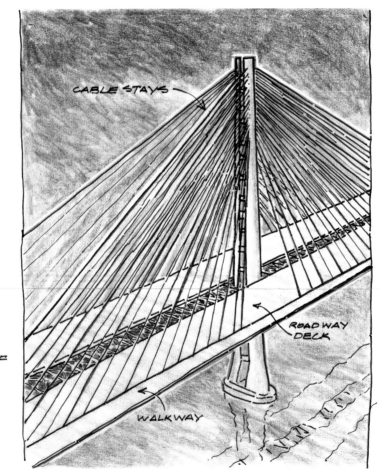

CABLE STAYS

ROADWAY DECK

CABLE-
STAYED
SINGLE
TOWER OF
THE MAIN
SPAN

WALKWAY

CABLE STAYS

ROADWAY
DECK

CABLE-STAYED DUAL TOWER OF THE MAIN
SPAN WITH VARIABLE DECK ELEVATIONS

But both of the competing designs would have been firsts: the first cable-stayed bridge to be built in California and the first self-anchored suspension bridge in the state and possibly the country. In more technical language, what MacDonald's design team called the first "mono-cable, deck-anchored, vehicular-carrying suspension bridge in the world" was chosen.

The difference in the proposed cable designs is notable. A self-anchored suspension span means it is not fixed to landfalls but to itself. The main cable is threaded underneath the west end of the span, with the two ends draped across the top of the tower before descending to the east end of the span, where they are anchored in vertical steel anchoring plates.

The effect would be like a webbed drape framing the main span, an application of cable never undertaken before. In the self-anchored suspension span, the cables extend from the anchoring plates westward and loop through the tower saddle, then cradle around the underside of the deck,

and return eastward to the tower saddle. Finally, they are secured to the anchoring plates on the eastern side. The cables are not secured to any land anchorage but use their own tension and shift of forces to maintain the tension necessary to support the bridge.

The proposed single tower of the self-anchored suspension bridge was actually four towers bound together by beams designed to move and deform, but not break, in an earthquake. The seismic impact was largely carried by the beams, which can be repaired or replaced. The tower would become the tallest self-anchored suspension support in the world.

Critics of the single-tower self-anchored suspension bridge, however, claimed the structure was unstable, the instability increasing when you make it self-anchoring or self-supporting. It is similar to anchoring a ship to itself, not to the sea bottom.

The alternate single-tower cable-stayed span is more traditional, with a set of radiating cables anchored directly from the tower to the bridge deck. The proposed cable-stayed design also has a two-legged tower with an elliptical cross section that would taper toward the top. Link beams join the two legs to provide extra seismic strength.

On June 24, 1998 the Metropolitan Transportation Commission voted eleven to one to approve the single-tower self-anchored suspension bridge design. Outgoing Oakland mayor Elihu Harris cast the single dissenting vote, calling the design "ugly." Incoming Mayor Jerry Brown called it a "freeway on stilts." Mayor Willie Brown of San Francisco, who first supported a northern alignment but then switched to a southern, was equally vocal. (A viaduct or skyway proposal later suggested by Jerry Brown suggested a flat roadway with no suspension span, a form as undistinguished as it is dull.)

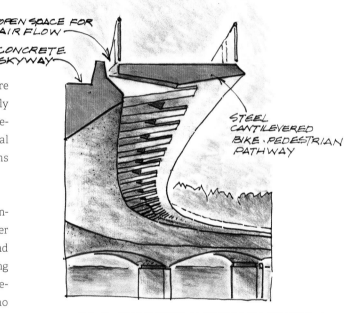

OPEN SPACE FOR AIR FLOW

CONCRETE SKYWAY

STEEL CANTILEVERED BIKE · PEDESTRIAN PATHWAY

BIKE · PEDESTRIAN PATHWAY SECTION ON THE SOUTH SIDE OF THE BRIDGE

THE SUSPENSION BRIDGES OF SAN FRANCISCO BAY

Opposite: Figure 49
Below: Figure 50

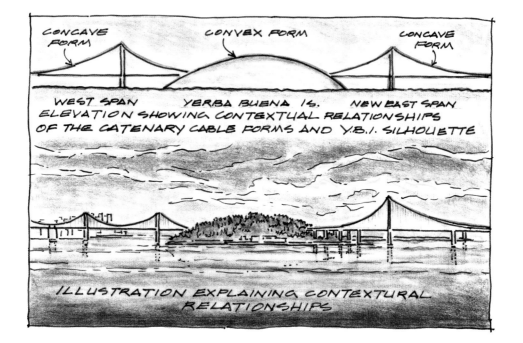

The navy again entered the fray, denying Caltrans access for its geological testing on Yerba Buena Island. They announced that an Environmental Impact Statement must first be prepared. Objections to the testing would go on for another year. A public initiative then expressed support for passenger rail service on the bridge. San Francisco and the navy again criticized Caltrans and their draft environmental statement, arguing that they had biased the process toward the selection of a northern alignment. Again, an impasse. Senator Barbara Boxer was then asked to intervene with the navy to allow soil testing.

A new governor, Gray Davis, asked for further advice on the alignment question, and a February 1999 hearing convened to hear new testimony from Berkeley professor Abolhassan Astaneh-Asl, who in 1992 had argued for retrofitting, not replacing, the existing bridge. The new bridge, he argued, would not withstand a major earthquake. Mayors Willie Brown and Jerry Brown asked Governor Davis to reconsider. But the delays were costing money,

and efforts to reopen the design process, now ten years after the earthquake, would be costly and take time, the governor replied. Davis then complained to the secretary of the navy about the delays, and an August 1999 meeting at the White House among contending parties only continued the impasse.

However, under a neglected law, the White House was able to transfer land from the navy to the highway administration, appropriating the land on Yerba Buena Island, so seismic work could be done on Interstate 80 where it crossed San Francisco Bay. Karen Skelton, who handled California issues for the Clinton White House, had become chief counsel for transportation secretary Rodney Slater and learned that the department could seize another agency's land to make improvements to the nation's highways. The navy was outclassed. By September 1999, the navy reluctantly gave its permission for soil testing.

The battle, however, was not over. Mayor Willie Brown continued to lobby the White House, meeting with John Podesta, White House chief of staff, and aides, arguing that a retrofit would be faster and less costly. To mediate the dispute, the Federal Highways Administration commissioned the Army Corps of Engineers to study the design of the East Span, although $70 million had already been spent on design and engineering for the new span. By May 2000, Caltrans had retrofitted 1,039 bridges, leaving 1,155 to complete. John Podesta then ordered the navy to transfer the disputed land to Caltrans, using the once-forgotten law empowering the Transportation Department to seize land to make improvements to the nation's highways. The decision, however, would not be made public for another five months, pending the result of the Corps of Engineers report.

The final choice by the MTC of the self-anchored suspension bridge immediately met with criticism from professionals and the public. Despite being recognized as a millennial project, a cultural undertaking to carry the region into the twenty-first century, critics complained that the design was not ambitious enough. Public debate and involvement, something of a San Francisco tradition, ran high. The scheme, representing only 15 percent of the entire East Span including its tower, was nonetheless the most visible and dramatic element of the entire project. The existing West Span would only be reinforced, while retaining its gray color and its undistinguished towers, called by *New York Times* architecture critic Herbert Muschamp "industrial, even martial poetry." Of the original East Span, the enclosed double-decker structure, he wrote that the form can't decide if it's a bridge or a tunnel.

The winning design was original: Its tapered tower of four pentagon-shaped steel columns and self-anchored main cable tied to the road deck forming its compression member, draping from the tower around the road deck, and looping back to the tower and down the short side to anchor under the roadway in a suspended steel

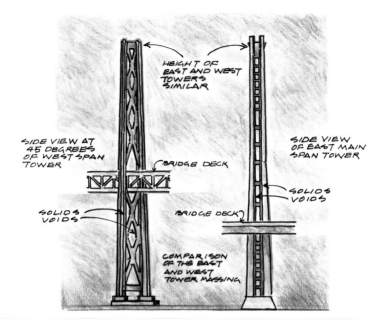

HEIGHT OF
EAST AND WEST
TOWERS
SIMILAR

SIDE VIEW AT
45 DEGREES
OF WEST SPAN
TOWER

BRIDGE DECK

SIDE VIEW
OF EAST MAIN
SPAN TOWER

SOLIDS
VOIDS

SOLIDS
VOIDS

BRIDGE DECK

COMPARISON
OF THE EAST
AND WEST
TOWER MASSING

FINAL SHAPE CHOSEN BECAUSE IT CUT DOWN THE MASS
OF THE TOWER LEGS AND IT WOULD APPEAR
SLENDER IN THE DAYLIGHT WITH THE SHADOWS
MOVING OVER THE ANGLED WALLS

AREAS FOR LINK BEAMS

SCHEME
ONE

SCHEME
TWO

SCHEME
THREE

FINAL
DESIGN

SCHEMATIC DESIGN PROCESS STUDIED FOR
THE MAIN TOWER SECTION

box close to Yerba Buena Island, was unique. Nothing like this had ever been tried before. Suspender cables attached the main cable to girders below the roadway. The three other runner-up designs—a cable-stayed structure with fan-shaped planes of cables running to the outside edges of the roadway, and two twin-towered designs with slim entry portals—were less aesthetic and functional. The weakest element in all the designs, however, was the undistinguished but necessary long causeway, although newly secured in up to three hundred feet of deep, stiff mud.

The tower is, in fact, the most critical design component and the keynote for the design vocabulary of the structure. It is the first thing you see as you approach the bridge, whether close or far. In terms of design, everything follows from its form, establishing the dominant aesthetic of the bridge. Emphasizing the incorporation of light in the structure is its open element, four pillars arching skyward that are actually separated but joined by link beams,

allowing light to stream through. As it narrows toward its top, sustaining the illusion of lightness and adding grace through perspective, the tower actually appears to be more elegant and less massive as it gets taller. The repeated use of solid then void establishes a rhythm of openness that furthers the lightness both in weight and actual mass of its form. It actually mimics the space between the legs on the existing West Span towers, joining the new span with the old *(Figure 51)*.

The towers, in fact, of the old and new spans are almost equal in height. The crossbeams and open space separating the east and west roadways of the bridge deck echo the solid/void cadence of the tower.

The pentagonal design of the tower legs sets the design vocabulary. From a vertical section looking downward, the tops of the four legs each have a pentagon-like shape, partly to reduce the sense of mass and increase the stability of the legs. This form is carried down to the pile caps supporting the skyway, repeated in the light poles and even the vertical

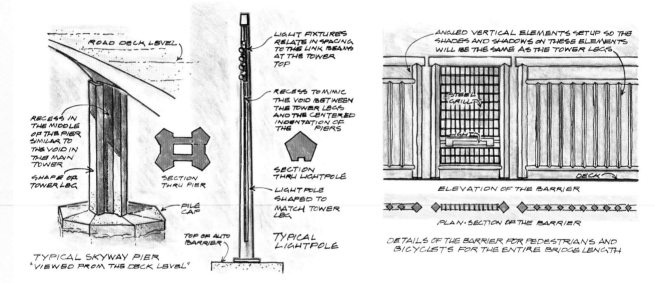

ROAD DECK LEVEL

LIGHT FIXTURES RELATE IN SPACING TO THE LINK BEAMS AT THE TOWER TOP

ANGLED VERTICAL ELEMENTS SETUP SO THE SHADES AND SHADOWS ON THESE ELEMENTS WILL BE THE SAME AS THE TOWER LEGS

RECESS IN THE MIDDLE OF THE PIER SIMILAR TO THE VOID IN THE MAIN TOWER

SECTION THRU PIER

SHAPE OF TOWER LEG

RECESS TO MIMIC THE VOID BETWEEN THE TOWER LEGS AND THE CENTERED INDENTATION OF THE PIERS

STEEL GRILL

LIGHT

DECK

PILE CAP

SECTION THRU LIGHTPOLE

LIGHTPOLE SHAPED TO MATCH TOWER LEG

TOP OF AUTO BARRIER

ELEVATION OF THE BARRIER

PLAN·SECTION OF THE BARRIER

TYPICAL SKYWAY PIER
"VIEWED FROM THE DECK LEVEL"

TYPICAL LIGHTPOLE

DETAILS OF THE BARRIER FOR PEDESTRIANS AND BICYCLISTS FOR THE ENTIRE BRIDGE LENGTH

Opposite: Figures 53 &54
Below: Figures 55 & 56

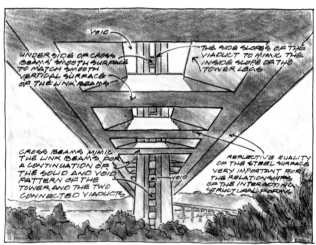

VIEW OF THE UNDERSIDE OF THE MAIN SPAN LOOKING
FROM YERBA BUENA ISLAND

Labels within figure:
VOID

UNDERSIDE OF CROSS
BEAMS' SMOOTH SURFACE
TO MATCH SMOOTH
VERTICAL SURFACE
OF THE LINK BEAMS

THE SIDE SLOPES OF THE
VIADUCT TO MIMIC THE
INSIDE SLOPE OF THE
TOWER LEGS

CROSS BEAMS MIMIC
THE LINK BEAMS FOR
A CONTINUATION OF
THE SOLID AND VOID
PATTERN OF THE
TOWER AND THE TWO
CONNECTED VIADUCTS

VOID

REFLECTIVE QUALITY
OF THE STEEL SURFACE
VERY IMPORTANT FOR
THE RELATIONSHIPS
OF THE INTERACTING
STRUCTURAL FORMS

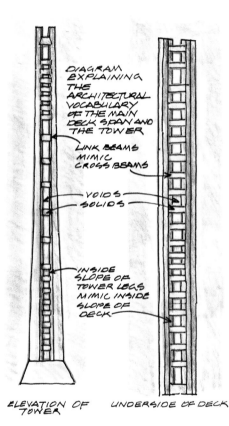

Labels within figure:
DIAGRAM
EXPLAINING
THE
ARCHITECTURAL
VOCABULARY
OF THE MAIN
DECK SPAN AND
THE TOWER

LINK BEAMS
MIMIC
CROSS BEAMS

VOIDS
SOLIDS

INSIDE
SLOPE OF
TOWER LEGS
MIMIC INSIDE
SLOPE OF
DECK

ELEVATION OF   UNDERSIDE OF DECK
TOWER

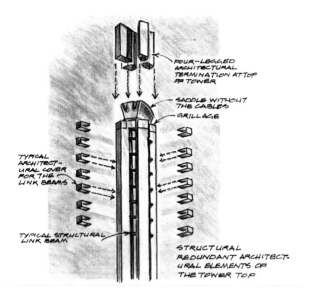

Four-legged architectural termination at top of tower

Saddle without the cables

Grillage

Typical architectural cover for the link beams

Typical structural link beam

Structural redundant architectural elements of the tower top

from the tower down to the tunnel in one direction and to the Oakland touchdown in the other. Key elements from the tower come off the structural system onto other features of the main span and the viaduct.

Occasional elements have only an aesthetic function, such as the covers over the sides of the link beams and the tower top (*Figure 57*). Their importance, however, is to sustain the unity of design throughout the bridge. The seven link beams with rounded façades at the tower top are repeated in the highest light poles, which range from thirty-two to sixty-four feet in height and repeat the one-third to two-thirds ratio, with the light fixtures one-third of the total length of the pole echoing the seven link beams that make up one-third of the new tower: The seven light fixtures form the top one-third, the remaining two-thirds of the pole are clear (*Figures 58–61*).

The unique, graduated lighting uses a new LED system that has a dramatic effect leading up to the illuminated tower itself. All the vertical elements of the East Span

pickets of the bridge railing (*Figures 52, 53 & 54*). Even the attachment of the crossbeams under the roadway deck echoes the sloping shape of the tower legs at the link beams (*Figures 55 & 56*).

Additionally, the light fixtures on the light poles pick up the multi-link beams at the tower top. Even the recess in the middle of the skyway pier and the light poles mimics the void in the main tower, while the pier cap itself is designed in a pentagon-like form. The bridge is consistent

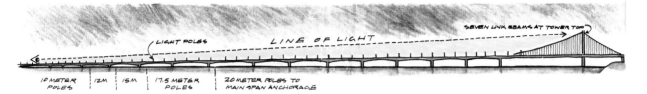

LIGHT POLES   LINE OF LIGHT   SEVEN LINK BEAMS AT TOWER TOP

10 METER POLES   12M   15M   17.5 METER POLES   20 METER POLES TO MAIN SPAN ANCHORAGE

CONTEXT OF LIGHTING SYSTEM OF THE SKYWAY POLES AND LIGHT FIXTURES IN RELATIONSHIP TO THE TOWER AND IT'S TOP

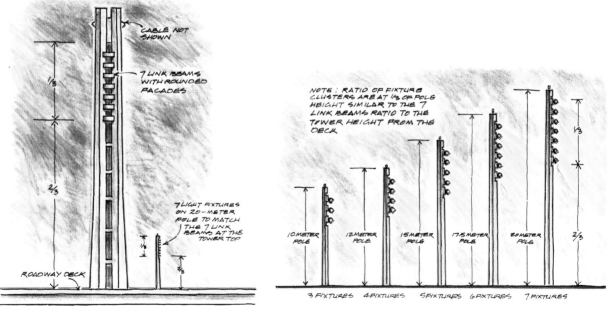

CABLE NOT SHOWN

7 LINK BEAMS WITH ROUNDED FACADES

1/3

2/3

7 LIGHT FIXTURES ON 20-METER POLE TO MATCH THE 7 LINK BEAMS AT THE TOWER TOP

1/3

2/3

ROADWAY DECK

RATIO RELATIONSHIP OF LIGHT POLES AND FIXTURES TO THE 1/3 RATIO ON THE TOWER

NOTE: RATIO OF FIXTURE CLUSTERS ARE AT 1/3 OF POLE HEIGHT SIMILAR TO THE 7 LINK BEAMS RATIO TO THE TOWER HEIGHT FROM THE DECK

10 METER POLE   12 METER POLE   15 METER POLE   17.5 METER POLE   20 METER POLE

1/3

2/3

3 FIXTURES   4 FIXTURES   5 FIXTURES   6 FIXTURES   7 FIXTURES

PROGRESSION OF LIGHT FIXTURES FROM THE EAST OAKLAND TOUCHDOWN TO THE TOWER OF THE MAIN SPAN

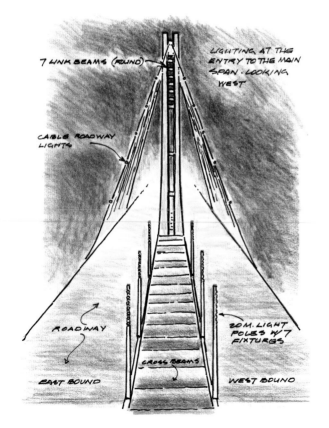

7 LINK BEAMS (ROUND)

LIGHTING AT THE ENTRY TO THE MAIN SPAN · LOOKING WEST

CABLE ROADWAY LIGHTS

ROADWAY

20 M. LIGHT POLES W/ 7 FIXTURES

CROSS BEAMS

EAST BOUND

WEST BOUND

(the tower, pier, light standards) emphasize clean, modern lines and intensify the light and shadow. Light poles and hand railings used to unify the different structural designs create a distinct, seamless appearance. The light poles actually vary in height and illuminated intensity to maintain a constant level of light on the roadway. With a line of light along the bridge all the way from Yerba Buena Island to Oakland, a white line now stretches across San Francisco Bay, extending a necklace of light running from the top of the tower and then along the roadway echoed on the side of the bridge. Utilizing the light poles and bike/pedestrian pathway, the bridge becomes a white line across the Bay *(Figures 62 & 63)*.

In fact, the ratio of fixture clusters are at one-third of pole height, similar to the seven link beams' ratio to the tower height from the deck. They grow taller as one moves westward from Oakland to the tower of the main span. The corners of the piers repeat the tower leg shape that even travels down to the footing of the pile

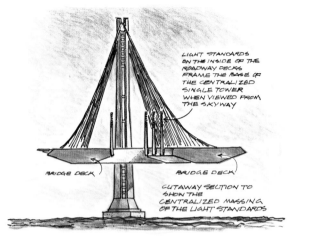

LIGHT STANDARDS ON THE INSIDE OF THE ROADWAY DECKS FRAME THE BASE OF THE CENTRALIZED SINGLE TOWER WHEN VIEWED FROM THE SKYWAY

BRIDGE DECK

BRIDGE DECK

CUTAWAY SECTION TO SHOW THE CENTRALIZED MASSING OF THE LIGHT STANDARDS

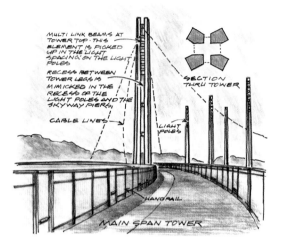

MULTI LINK BEAMS AT TOWER TOP. THIS ELEMENT IS PICKED UP IN THE LIGHT SPACING ON THE LIGHT POLES

RECESS BETWEEN TOWER LEGS IS MIMICKED IN THE RECESS OF THE LIGHT POLES AND THE SKYWAY PIERS

CABLE LINES

SECTION THRU TOWER

LIGHT POLES

HANDRAIL

MAIN SPAN TOWER

caps, while the underside of the bridge deck picks up the recesses seen on the link beams and piles viewed from the underside.

Color also works to unite the elements of the bridge, the white tower repeated on the painted bike/pedestrian pathway. Painted white, the steel establishes consistency of the main deck with the viaduct to give the sense of a continuous horizontal form as one looks at the bridge, the water reflecting a kind of intense but mottled light on the white underside of the bridge, which will also echo the white bike path. There should be no visual or aesthetic separation of the main span from the skyway; it should

be as one. The white painted bike path edge of the skyway also picks up the reflection of the sunlight, the overall white of the tower and bridge reflecting the highly visible white cranes of Oakland (*Figure 64*).

The placement of the main span's bridge deck is particularly important, as it sits two-thirds below the tower top and one-third above the water surface. This principle of thirds, a feature of classical architectural proportions, visually intrigues the eye and pulls it to the form, rather than the roadway simply bisecting the distance between the top of the tower and the water. The existing West Span shows how uninteresting this can be, as the

WHITE CRANES OF OAKLAND

·SOUTH· ELEVATION OF BRIDGE STRUCTURES ·FROM TUNNEL TO TOUCHDOWN·

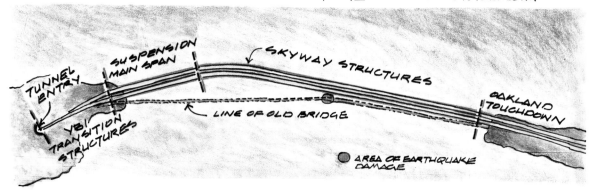

MAP OF BRIDGE STRUCTURES NOMENCLATURE

YEONGJONG BRIDGE · KOREA 2000
WORLD'S LONGEST SELF-ANCHORED BRIDGE
MAIN SPAN 550 METERS

KONOHANA BRIDGE · OSAKA · JAPAN
1990 · 300-METER SPAN · SINGLE
SELF-ANCHORING CABLE

roadway bisects the four existing towers. The very imbalance of the two-thirds to one-third ratio creates visual interest. Photographers understand this by never shooting with the horizon in the middle of their pictures. It is always above or below the middle of the frame *(Figure 65)*.

The placement of the light fixtures on the light poles also reflects this ratio. As you move closer to the tower, the light poles change in size, increasing in height as one gets closer to the tower. The idea is to create a crescendo effect as one gets nearer the dominating tower form. And

as there are seven link beams in each section of the tower, there are seven lights on each ascending pole to create a growing line of light.

The cables of the bridge actually embrace the light poles. Centered in the inside edge of the decks and not on the outside of the skyway, they allow for greater visibility for drivers, pedestrians, and bicyclists while creating a gateway effect leading to, or leaving from, the tower. Marker lights running up the cables to the tower top continue the markers on the roadway light poles, further creating a dramatic lightscape. Even the bike path handrails have a sense of passage through the pentagon system, repeating the design in its handrails and posts. The vertical pickets turned at a forty-five-degree angle also give the impression of a pentagon. This system of pentagonal forms collectively tie the major and minor elements of the bridge and the skyway together to create an artistic whole, the complete opposite of the original East Span with its contrasting form of the deck trusses and then cantilevered main span trusses, nothing connected. In the new form, everything unites.

To build a self-anchored suspension bridge, you actually need to build two, one a falsework to support the main form as it is built. With a conventional suspension bridge, you construct the towers first. Main cables are then hung between them and the deck is attached to the cables. A self-anchored suspension bridge reverses the process. Because the suspension cables are anchored to the deck rather than in the ground at either end of the bridge, the deck must be placed high above the water on a temporary edifice known as a falsework. One needs a scaffolding of trusses underneath the box girders (the deck) for support. The scaffolding is required to hold the deck as it is being built out. Then, the cables attaching it to the tower are connected. In essence, you build two bridges, the first a temporary structure to support the roadway before it is cabled, which is then removed.

During World War II, self-anchoring spans were popular for small bridges because they were relatively cheap and easy to construct. Before seismic engineering was developed, ground tremors caused by nearby bombing in WWII occasionally resulted in collapse. Today, however, they exist throughout the world, two examples being the Yeongjong Grand Bridge in Korea and the Konohana Bridge in Osaka, Japan *(Figures 66 & 67)*. When completed, the Bay Bridge's East Span will exceed them in length.

6 BUILDING THE WHITE SPAN

After the final design selection and approval by the Metropolitan Transportation Commission and the legislature, an invitation for bids to build the East Span Tower went out, the $1.04 billion contract for the skyway from Oakland to the self-anchored suspension span previously awarded in January 2002. Only one bid had been received for the self-anchored suspension span and tower, in May 2004, and it was for $1.8 billion, more than twice the $733 million estimate. The state declined to contract with the bidder after the legislature did not approve funding by the deadline of September 30, 2004, at the same time a review of cost estimates recorded increasingly higher costs. Disappointingly, the administration of Governor Arnold Schwarzenegger—elected in 2003—reported in August 2004 that the estimated cost of rebuilding the East Span had gone up to $5.1 billion from an earlier estimate of $2.6 billion. They blamed at least half the cost increase on the self-anchored suspension portion of the bridge. Other cost increases included the skyway section, up almost half a billion more than when the contract was signed.

Explanations for the higher costs were attributed to labor, material, insurance, and the price of steel and concrete accelerated by China's preparations for the Olympics, compounded by post-9/11 safety, security, and insurance concerns worldwide. Other costs included a seismic retrofit of BART's transbay underwater tube. A new risk assessment of the Bay Bridge project by the U.S. Federal Highway Administration added more delay, although work on the foundations for the skyway and for the suspension span at Yerba Buena Island, which began in the early 2000s, continued.

Unsure of its next step in the face of major cost overruns, Caltrans held several meetings over the possibility of a bridge redesign, with six new proposals, including the self-anchored suspension span, a cable-stayed design, and an extended viaduct. An independent review team endorsed a cable-stayed design because it would be simpler to construct and present fewer risks of schedule delays than the self-anchored suspension span and could possibly save over $600 million. The U.S. Peer Review Team complicated the matter, however, by arguing that the risks of alienating the public would be

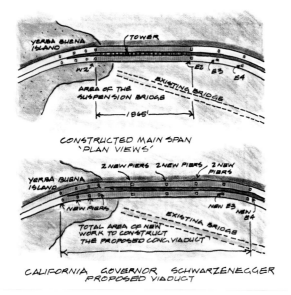

CONSTRUCTED MAIN SPAN
'PLAN VIEWS'

CALIFORNIA GOVERNOR SCHWARZENEGGER
PROPOSED VIADUCT

higher with a skyway design (essentially a freeway on stilts) and a costlier cable-stayed alternative. But in terms of cost overruns and delays, the self-anchored form was the riskiest.

Caltrans finally recommended two plans to Governor Schwarzenegger: either rebid a modified self-anchored plan or extend the concrete viaduct from Oakland to Yerba Buena Island. In October 2004, the governor halted the self-anchored plan and endorsed the concrete viaduct,

extending it another 1.1 miles to supposedly save up to $500 million, while avoiding the technical complexity of the self-anchored design. (Seven years earlier, this same proposal had been rejected.) A graph from May 2004 vividly showed the ballooning costs from the 1999 estimate for the entire East Span at $1.3 billion to the 2004 estimate of $4 billion, which by 2011 had become $6.3 billion.

Schwarzenegger's supposed cost-saving plan required six additional foundations, which meant reducing the open channel—a secondary entry for the bay—from 1,200 to 750 feet, considered a cautionary, minimal width for ships to pass each other; 1,000 feet was needed for two-way traffic (*Figures 68 & 69*).

At this time, a design by Frank Lloyd Wright, prepared in 1949 for a southern crossing of the bay, resurfaced (*Figure 70*). A unique butterfly bridge, it had a roadway dividing and curving around an oasis of grass and trees at mid-span. The reinforced concrete structure rested on

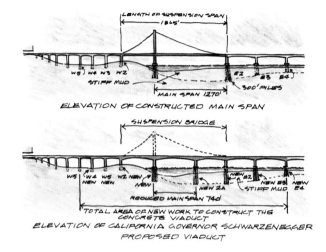

ELEVATION OF CONSTRUCTED MAIN SPAN

ELEVATION OF CALIFORNIA GOVERNOR SCHWARZENEGGER PROPOSED VIADUCT

a series of hollow piers sunk into the bay and called by Wright "tap-roots." Large, hollow, curved slabs sprung out like huge fans and spread eighty feet on each side of a pier supporting the seventy-foot-wide roadway, carrying six lanes of traffic and two pedestrian walks. Two twin arches two-hundred feet above the main channel split the traffic in different directions but were connected at their crowns by a garden. The "Butterfly Wing Bridge" was supposed to connect San Francisco at what is currently Chavez and Third Street to an eastern point on Bay Farm Island just north of the Oakland Airport. Essentially a low viaduct until it reached the shipping channel, rising up to an archway spanning 2,000 feet, the "Butterfly Wing" was never built, although construction of a scale model was completed in 1951.

By February 2005, lawmakers, the public, and MacDonald voiced objections to the proposed change, arguing that the cost of the causeway extension was not known and

that the obligation for the contract for new foundations of $175 million was still in place, even if the pillars might not be compatible with the proposed viaduct extension. Costs at this time were also soaring, the latest estimate $5.1 billion for the East Span. A viaduct extension would also take years of revised design work, as well as an extension of the environmental study, and likely cost as much, if not more, than the current plan because of contractors who have had to stop work and be paid. The skyway would also have to be reengineered to fit with any redeveloped span and require six more foundations placed at least three hundred feet deep. The steel required to hold up the box girder could actually cost more than the steel in the current tower and deck. A redesign would also delay the project for three years. A viaduct the entire length would also contrast with the harmony of the bridges in the Bay Area.

By contrast, the single-tower span was ready to build. Although the governor had decided to suspend the tower design, others pointed out that it was part of the state law authorizing and funding the project. New legislation would be needed to change it. Testifying in Sacramento, MacDonald and his assistant Caspar Mol argued that their single-tower self-anchored suspension bridge design, while possibly costlier, was safer than a viaduct. Politicians again divided, with Democrats and elected officials from the Bay Area supporting the original and approved East Span design, the governor and down-state Republicans opposing it.

By August, the tower was back on track. Finances, the main reason to initially halt construction of the tower, became the method to solve the impasse. A new agreement to complete construction and financing was reached in late June, approved a few days later. The state would contribute an additional $630 million to the total retrofit program, with $300 million going to the demolition of the old bridge. An increase in tolls would handle much of the refinancing, with tolls going up by $1 on seven state-owned bridges in 2007. This compromise allowed Caltrans to advertise for bids on the self-anchored suspension tower and East Span .

In March of the following year, two bids were unsealed, both close to the engineers' estimates of what had become a $6.3 billion project, with $1.4 billion for the self-anchored suspension section. The construction contract also had early-completion incentives of $50,000 a day. The Schwarzenegger delay, however, meant a year lost in the process and increased costs, since steel and

FRANK LLOYD WRIGHT'S 1949
BUTTERFLY WING BRIDGE FOR THE
SAN FRANCISCO BAY

other construction charges increased. Controversy over using foreign steel fabricated in China then erupted, but since no federal funds were used in the self-anchored structure, no "buy American" requirement existed.

The construction of the bridge decks occurred 6,500 miles away in Shanghai in a huge facility and involved three thousand workers. Barges delivered the last of the giant steel frames for the roadbed in August 2011, when a giant crane then lifted them into position. Assembly work, as well as the pouring of the concrete road surface, occurred in Oakland. Using imported steel saved an estimated

$450 million on the roadway and tower with its four vertical sections, each section weighing on average 700 tons. The total tower weight was calculated to be 14,850 tons. Individual road deck panels weighed 2,200 tons. Temporary towers supported the 34,000-ton bridge until installation of the 4,600-foot-long single cable took place.

Assembly required the largest construction hydraulic hammers and cranes in the world, as well as the world's largest precast concrete segments (three stories tall) for the skywalk deck (Figures 71 & 72).

The entire span contains almost 200 million pounds of structural steel, 120 million pounds of reinforcing steel,

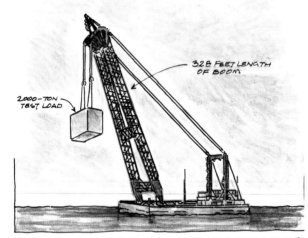

2000-TON
TEST LOAD

328 FEET LENGTH
OF BOOM

BAY BRIDGE · LEFT · COAST LIFTER · LARGEST BARGE CRANE
ON THE WEST COAST BEING TESTED WITH A 2000-TON LOAD

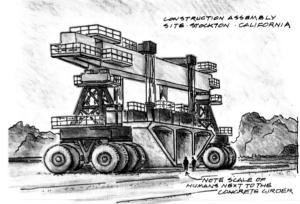

CONSTRUCTION ASSEMBLY
SITE · STOCKTON · CALIFORNIA

NOTE SCALE OF
HUMANS NEXT TO THE
CONCRETE GIRDER

CUSTOM-MADE LIFT MACHINERY FOR MOVING
THE CONCRETE SEGMENTAL GIRDERS

Opposite: Figures 71 & 72
Below: Figure 73

SOUTH ELEVATION

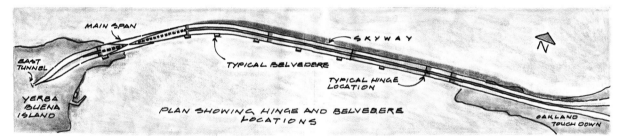

MAIN SPAN

SKYWAY

EAST TUNNEL

TYPICAL BELVEDERE

TYPICAL HINGE LOCATION

YERBA BUENA ISLAND

PLAN SHOWING HINGE AND BELVEDERE LOCATIONS

OAKLAND TOUCH DOWN

N

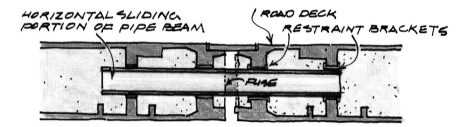

HORIZONTAL SLIDING
PORTION OF PIPE BEAM

ROAD DECK
RESTRAINT BRACKETS

PIPE

LONGITUDINAL SECTION THRU PIPE BEAM AT HINGE

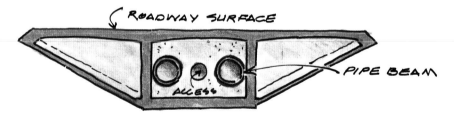

ROADWAY SURFACE

PIPE BEAM

ACCESS

CROSS SECTION THRU PIPE BEAM AND VIADUCT

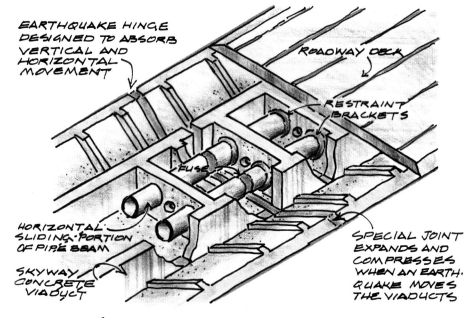

EARTHQUAKE HINGE
DESIGNED TO ABSORB
VERTICAL AND
HORIZONTAL
MOVEMENT

ROADWAY DECK

RESTRAINT
BRACKETS

FUSE

HORIZONTAL
SLIDING·PORTION
OF PIPE BEAM

SKYWAY
CONCRETE
VIADUCT

SPECIAL JOINT
EXPANDS AND
COMPRESSES
WHEN AN EARTH·
QUAKE MOVES
THE VIADUCTS

SPECIAL PIPE BEAMS AT KEY
STRESS POINTS ALONG THE SKYWAY

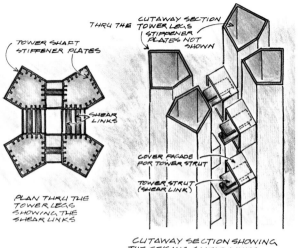

TOWER SHAFT
STIFFENER PLATES

THRU THE
CUTAWAY SECTION
TOWER LEGS
STIFFENER
PLATES NOT
SHOWN

SHEAR
LINKS

PLAN THRU THE
TOWER LEGS
SHOWING THE
SHEAR LINKS

COVER FACADE
FOR TOWER STRUT

TOWER STRUT
(SHEAR LINK)

CUTAWAY SECTION SHOWING
THE SEISMIC SHEAR LINKS

and approximately 450 thousand cubic yards of concrete. The asymmetrical feature of the suspension strand has a longer forward span, east of the tower, than back span, which offers a more gradual transition from the progressively sloping skyway.

To insure seismic safety on the skyway, an advanced hinge-beam system of steel tubes six and a half feet in diameter placed between each segment of the deck was installed (Figures 73, 74 & 75).

The steel tubes allow the deck segments to slide during an earthquake and will absorb earthquake energy. So, too, will the East Span single tower. Embedded in rock,

the single tower with its four separate legs connected by link beams will take the impact from an earthquake and prevent damage to the tower legs (Figures 76 & 77). If one of the link beams sustains damage, the other will keep the bridge standing. The steel link beams will also shear off if the movement becomes too violent, allowing the other legs to stand. The self-anchored form in which the suspender cables are anchored to the deck itself was necessary because the geological conditions of the bay could not support cable anchor foundations where the new East Span is located. Consequently, a single suspension cable wraps over the tower and underneath the western end of the span before wrapping over the tower again to anchor in both roadway decks at the eastern end (Figures 78–82).

A single, nearly one-mile-long cable 2.6 feet wide wraps around the west end before returning back up the tower to anchor back in the east end. It is the longest looped suspension cable in any bridge in the world, with more than seventeen thousand 5-millimeter wires, each of which can supposedly support the weight of a military Hummer. The cable itself weighs 5,291 tons (nearly 10.6 million pounds).

Suspenders connect diagonally from the cable that crosses over the roadway to the outside edges of the deck. New seismic monitors will also be part of the bridge at key locations to record displacements and forces generated by earthquakes or other events. Assisting in the construction was a unique Pier Table to offer additional stabilization *(Figure 83)*.

The self-anchored suspension span is truly international, incorporating parts from around the world: cable bands from England, saddles from Japan, the main cable itself from China, suspender cables from Korea, the steel from China, and the concrete from the United States. Many of the construction drawings were done in Canada *(Figure 85)*.

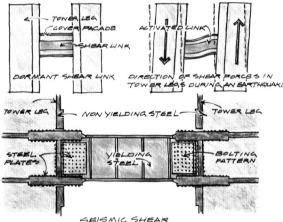

BICYCLE/PEDESTRIAN PATHWAY

SADDLE

MAIN CABLE

SADDLE

SADDLES

MAIN CABLE

CONTINUOUS MAIN CABLE AND THE FOUR
SADDLE LOCATIONS

SADDLE

SLOTS FOR
PRE-ASSEMBLED
CABLE BUNDLES

NOTE SCALE OF
HUMANS TO THE
SADDLE

TEMPORARY
FLOORING

LOOKING INTO THE SADDLE AT DECK EDGE ON THE
WEST END OF THE MAIN SPAN

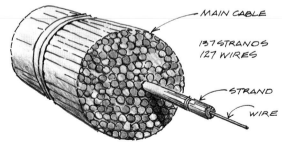

MAIN CABLE

137 STRANDS
127 WIRES

STRAND

WIRE

TYPICAL SECTION THRU
THE MAIN CABLE

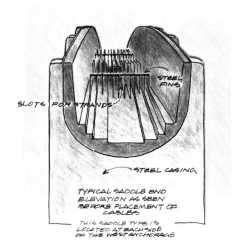

SLOTS FOR STRANDS

STEEL FINS

STEEL CASING

TYPICAL SADDLE END
ELEVATION AS SEEN
BEFORE PLACEMENT OF
CABLES

THIS SADDLE TYPE IS
LOCATED AT EACH SIDE
OF THE WEST ANCHORAGE

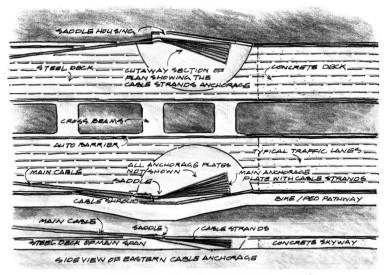

SADDLE HOUSING

STEEL DECK

CUTAWAY SECTION OF
PLAN SHOWING THE
CABLE STRANDS ANCHORAGE

CONCRETE DECK

CROSS BEAMS

AUTO BARRIER

TYPICAL TRAFFIC LANES

MAIN CABLE

ALL ANCHORAGE PLATES
NOT SHOWN

SADDLE

CABLE SHROUD

MAIN ANCHORAGE
PLATE WITH CABLE STRANDS

BIKE / PED PATHWAY

MAIN CABLE

SADDLE

CABLE STRANDS

STEEL DECK OF MAIN SPAN

CONCRETE SKYWAY

SIDE VIEW OF EASTERN CABLE ANCHORAGE

EASTERN CABLE ANCHORAGE BLOCK
'SELF ANCHORED'

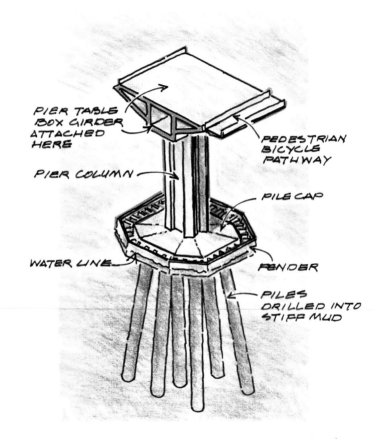

PIER TABLE
BOX GIRDER
ATTACHED
HERE

PEDESTRIAN
BICYCLE
PATHWAY

PIER COLUMN

PILE CAP

WATER LINE

FENDER

PILES
DRILLED INTO
STIFF MUD

TYPICAL PIER TABLE CONSTRUCTION
FOR THE SKYWAY

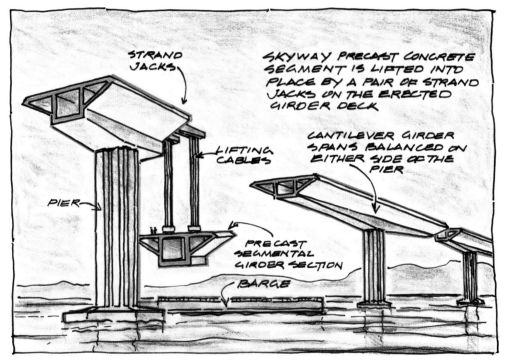

STRAND JACKS

SKYWAY PRECAST CONCRETE SEGMENT IS LIFTED INTO PLACE BY A PAIR OF STRAND JACKS ON THE ERECTED GIRDER DECK

LIFTING CABLES

CANTILEVER GIRDER SPANS BALANCED ON EITHER SIDE OF THE PIER

PIER

PRECAST SEGMENTAL GIRDER SECTION

BARGE

SEGMENTAL CONSTRUCTION OF SKYWAY STRUCTURE

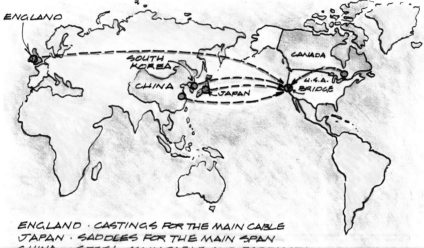

ENGLAND · CASTINGS FOR THE MAIN CABLE
JAPAN · SADDLES FOR THE MAIN SPAN
CHINA · · STEEL · MAIN CABLE AND FABRICATION
SOUTH KOREA · SUSPENDER CABLES
CANADA · CONSTRUCTION DRAWINGS

SOURCES OF INTERNATIONAL PARTICIPATION IN THE
CONSTRUCTION OF THE NEW EAST BRIDGE SPANS

Opposite: Figure 85

Among the unique features of the East Span are its length, its dual roadway with five lanes on each side with a fifty-foot open space separation in between the roadway decks, and the presence of two service lanes for break-downs and service on each side. Another feature is a bike and pedestrian path on the south side of the bridge, which will make it possible to circle the entire bay, once a new bike addition occurs on the West Span. The proposed four hundred-mile multi-use trail is near completion at this writing. Seven viewing platforms on the East Span allow cyclists and pedestrians to enjoy sweeping views.

Once construction started, the public was enthralled. Commuters could daily watch the progress as they traveled on the existing Bay Bridge or follow developments on a special Bay Bridge website with video links. The segmental construction with the dramatic lifting of roadway sections was especially fascinating (Figure 84). The segmental concrete box girders, actually produced in Stockton, California, were barged down the Sacramento River to the bridge. And once the tower began to ascend in 2009, it was a huge news story, with the construction made even more compelling by a new anxiety: a bridge failure on October 27, 2009, when a portion of the repaired, original East Span failed because several eye-bars snapped. The cause may have been metal-on-metal vibration from bridge traffic and wind gusts of up to fifty-five mph triggering the failure of one rod that broke off, which then led to the metal section crashing down. The actual pieces that broke off were a saddle, crossbars, and two tension rods. Three vehicles were either struck or hit by the fallen debris, although there were no injuries.

During the 2009 Labor Day weekend, there was a further closure for a portion of the replacement section, when a major crack was found in another eyebar. Working in tandem with the retrofit, Caltrans and its contractors and subcon-tractors were able to design, engineer, fabricate, and install the pieces required to repair the bridge, delaying its planned reopening on November 2 by only one and a half hours.

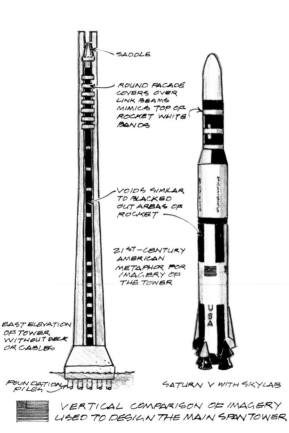

SADDLE

ROUND FACADE COVERS OVER LINK BEAMS MIMICS TOP OF ROCKET WHITE BANDS

VOIDS SIMILAR TO BLACKED OUT AREAS OF ROCKET

21ST-CENTURY AMERICAN METAPHOR FOR IMAGERY OF THE TOWER

EAST ELEVATION OF TOWER WITHOUT DECK OR CABLES

FOUNDATION PILES

SATURN V WITH SKYLAB

VERTICAL COMPARISON OF IMAGERY USED TO DESIGN THE MAIN SPAN TOWER

LIFTING CRANE

SADDLE

GRILLAGE

PLACEMENT OF THE SADDLE ON THE TOWER TOP GRILLAGE

Never had anything so large as the new East Span been constructed in the state. The tower itself resembled a *Saturn V* rocket, the tallest, heaviest, and most powerful launch vehicle ever built *(Figure 86)*.

For MacDonald, the *Saturn V* was an acknowledged prototype of the tower, a twenty-first-century metaphor for the structure. Television shows and the Bay Bridge website recorded the tower section lifts, which captured public interest, the *San Francisco Chronicle* reporting in breath-taking style the installation of the fourth section of the new tower as ascending "white sticks of steel." The 105.6-foot, 500-ton section had to be lifted from a barge and then gently raised into place more than forty stories above the bay. The first section went up in July, the second in October, the third in December 2010, and the fourth in March 2011. Only the grillage, the final section with the saddle, remained, finally installed in May 2011.

The ascent of the 525-foot tower in five stages— capped by placing the double-cable saddle, the world's largest, on the top—was reported by the press in detail *(Figure 87)*. The lift of the fourth stage of the tower meant that the tower was 91 percent complete, reaching 480 feet, each section weighing nearly a million pounds. The vertical presence of the tower was now visible throughout the Bay Area, its white form dominating the eastern skyline. And with the skyway completed, the construction of the main span decks was ready to begin.

7 FALCONS & FISH: THE ENVIRONMENT

One of the greatest concerns in building the East Span was the fragile ecology of the bay and the environmental impact of construction on the habitat, from sea life to the nesting cormorants under the deck of the old bridge *(Figures 88–94)*.

The presence of peregrine falcons and marine animals like the green sturgeon intensified the need to sustain the environmental balance as much as possible during and after construction. Caltrans, along with various environmental groups, carefully studied the biology of the bay and produced a comprehensive plan, formally an environmental-impact statement, released in 2001 after four years of study, to mitigate damage. Its acceptance by the legislature and environmental groups meant that bids could begin for the actual construction of the skyway and the self-anchored suspension bridge. The report included innovative ways to accommodate the natural life and public assets of the region, which naturally posed challenges to the bridge designers and engineers.

Five areas became the focus of the biological mitigation effort, beginning with water quality. A combination of erosion-control methods, including silt fencing and slope stabilization, prevented pollution of the bay. Backup systems for fuel storage and platforms to control mud during excavation were also built. Special basins collected and treated stormwater runoff from the 143 acres of the Bay Bridge approach, improving the water and habitat quality in the Emeryville Crescent Marsh and the San Francisco Bay.

To ensure the safety and well-being of the many birds in the area, special platforms were built under the new East Span to provide nesting places for the double-crested cormorants that roost under the Bay Bridge. Observers noted that eight hundred pairs of the birds (one of the largest colonies in California) lived under the original Bay Bridge. New ledges and platforms were also constructed on the East Span to accommodate the fishing birds that dive from the bridge for their food. Crews also constructed a 500-square-foot island to serve as a habitat for roosting shorebirds, including the snowy egret and the ruddy turnstone *(Figures 95–101)*.

Marine mammals required special attention, especially during the driving of the 259 large-diameter piles—hammered at angles ("battered" in engineering terms) for maximum strength to depths of up to 310 feet—into the

CALIFORNIA SEA LION

CALIFORNIA GRAY WHALE

HARBOR PORPOISE

HARBOR SEAL

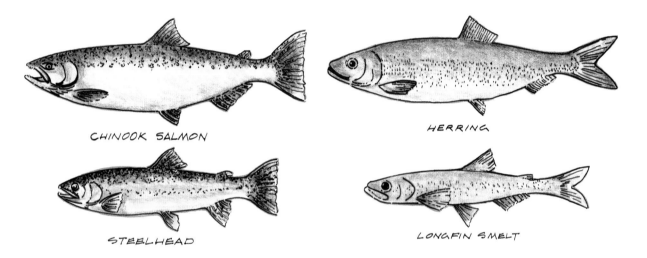

CHINOOK SALMON

HERRING

STEELHEAD

LONGFIN SMELT

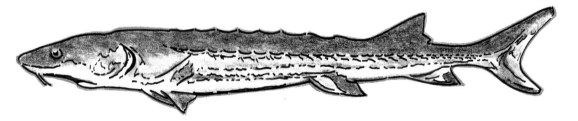

GREEN STURGEON

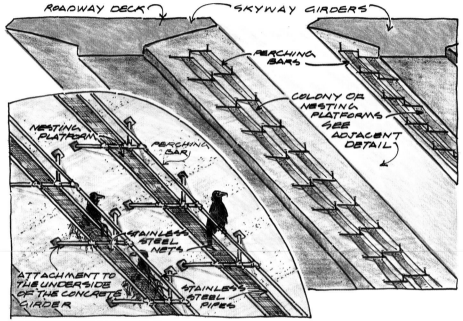

ROADWAY DECK

SKYWAY GIRDERS

PERCHING BARS

COLONY OF NESTING PLATFORMS SEE ADJACENT DETAIL

NESTING PLATFORM

PERCHING BAR

STAINLESS STEEL NETS

ATTACHMENT TO THE UNDERSIDE OF THE CONCRETE GIRDER

STAINLESS STEEL PIPES

CONSTRUCTION DETAIL OF THE NESTING PLATFORMS

CORMORANTS NESTING PLATFORMS ON THE UNDERSIDE OF THE EASTERN END OF THE SKYWAY

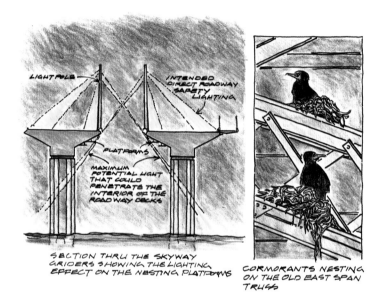

SECTION THRU THE SKYWAY GRIDERS SHOWING THE LIGHTING EFFECT ON THE NESTING PLATFORMS

CORMORANTS NESTING ON THE OLD EAST SPAN TRUSS

bay floor, which were needed for the skyway and other deep-water supports. (The existing East Span consisted of only eighty-five-foot-long timber piles.) Hydro-acoustic monitoring examined the impact of the underwater pressure on fish, with the discovery that the pressure from sound waves from the hydraulic pile-driving hammers collapsed the bladders of many fish. As a result, innovative bubble curtains were built around the pile drivers to mitigate the sound and reduce the pressure in the

water that had shocked the fish. Weekly bird monitoring also occurred, continuing throughout the construction process. Monitors reported a diverse population, from Canada geese to Western gulls.

Eelgrass, an aquatic plant critical to the bay's ecosystem, is an important marine habitat that improves water quality through its collecting and filtering organic matter and sediments. Remote turbidity monitors in the bay that monitor changes in the water quality that could

WESTERN GULL

DOUBLE-CRESTED CORMORANT

CALIFORNIA BROWN PELICAN

AMERICAN PEREGRINE FALCON

CALIFORNIA LEAST TERN

impact the eelgrass were set up throughout the area. At the Oakland touchdown, environmental staff installed a curtain to control turbidity and protect eelgrass beds. Experimental beds of eelgrass have also been transplanted to study its growth and survival. And throughout the construction area in the bay and shore, fences, signs, and buoys note and protect sensitive areas *(Figure 102).*

Another of the most important concerns was the disposal of dredged material from the bay, anticipated to be 450,000 cubic yards. This mix of sand and dirt, either dredged or churned up by the construction, went to deep-ocean dumping sites or other sections of the bay showing erosion.

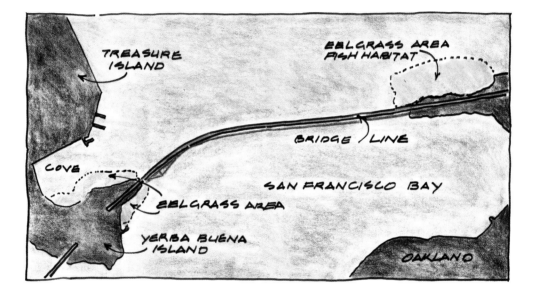

MAP SHOWING AREAS OF EELGRASS
"FISH HABITAT"

8 THE
FUTURE

The new Bay Bridge is a bridge for the twenty-first century. Its form, structural elements, earthquake stability, and new technology allowing for flexibility and movement, plus its revolutionary construction methods, have resulted in the most innovative and longest self-anchored suspension bridge in the world. Its inclusion of pedestrians and bike paths will make it possible to stroll or cycle over the entire Bay Area. The vistas, experienced by drivers and pedestrians, will be extraordinary. From its unique lighting to its dual roadway, it will be a signature form *(Figures 103, 104 & 105)*.

The proposed bike-pedestrian paths linking the East Span to the West also suggest the new character of travel. The lanes will go over the tunnel and around Yerba Buena Island before joining both sides of the upper deck of the West Span. It will then follow the roadway down into San Francisco to street level. The meeting of the new bridge and existing tunnel entry shows how the deck and pylons announce the transition to the existing art deco form *(Figures 106–111)*.

But the double suspension bridge making up the West Span (which underwent a five-year seismic upgrade beginning in 1999) should not be forgotten. Retrofitting consisted of reinforcing and stiffening the original span with new steel plates and replacing the original half million rivets with one million high-strength bolts. Seventeen million pounds of structural steel were added, as well as new bracing under the deck. Piles were encased in heavy concrete jackets, while additional anchor bolts were installed to fasten the tower legs to their pedestals. Ninety-six dampers were also added to serve as shock absorbers. The retrofit involved a thousand workers, often round-the-clock.

Further changes involving the Bay Bridge included the June 2011 approval by the San Francisco Board of Supervisors for a new development on Treasure Island of 403 acres on a former naval base adjacent to Yerba Buena Island. With a unanimous vote, a massive new neighborhood proposal worth $1.5 billion, for 19,000 new residents with a number of high-rises, one as tall as 450 feet, was approved. A new ferry terminal as well as a school will be built. The concept is to

TIME CLOCK
DATE OF DAY
TEMPERATURE

TIME CLOCK
DATE OF DAY
YEAR

SIGN STRUCTURE

OAKLAND

EAST BOUND TRAFFIC

PURPOSED OAKLAND GATEWAY PYLONS
NOT BUILT

BAY BRIDGE

EXISTING AND PROPOSED
BICYCLE TRAIL FOR
PASSAGE OVER THE BAY
BRIDGE

SAN FRANCISCO BAY

BICYCLE
TRAIL

OAKLAND   EMERYVILLE   BERKELEY   ALBANY   EL CERRITO

EASTSHORE STATE PARK
MAIN BAY BRIDGE BICYCLE TRAIL

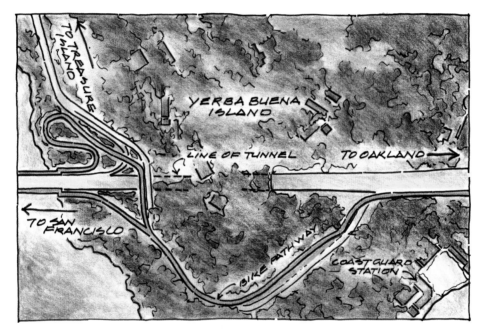

PROPOSED BIKE · PEDESTRIAN PATHWAY AROUND
YERBA BUENA ISLAND

PROPOSED BIKE·PEDESTRIAN PATHWAY OVER
TUNNEL ENTRY ON YERBA BUENA ISLAND

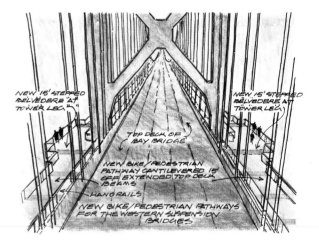

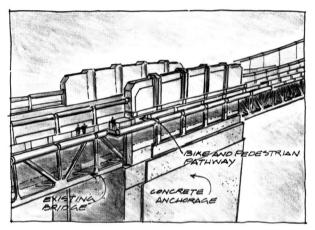

PROPOSED BIKE AND PEDESTRIAN PATHWAY AT
CENTRAL ANCHORAGE OF THE WEST SPANS OF
THE BAY BRIDGE

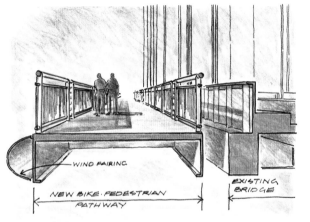

SECTION THRU NEW PATHWAY ON
THE WEST SPANS OF THE BAY BRIDGE

THE FUTURE

YERBA BUENA ISLAND

TUNNEL ENTRY 1930'S ART DECO ARCHITECTURE

NEW BRIDGE HEADS DESIGNED TO MATCH FORMS OF THE TUNNEL ENTRY

END OF 21ST CENTURY EAST SPAN. THIS MARKS AND SHOWS THE TRANSITION BETWEEN THE OLD AND NEW ARCHITECTURAL AND STRUCTURAL FORMS

THIS DRAWING ILLUSTRATES THE MEETING OF THE NEW BRIDGE AND THE INTRODUCTION OF THE ART DECO ARCHITECTURE

TOLL BOOTHS

NEW TOLL ADMINISTRATION BUILDING AND CANOPY · 2013

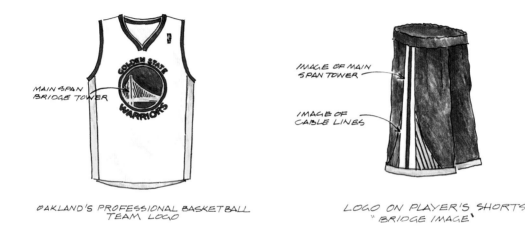

MAIN SPAN
BRIDGE TOWER

IMAGE OF MAIN
SPAN TOWER

IMAGE OF
CABLE LINES

OAKLAND'S PROFESSIONAL BASKETBALL
TEAM LOGO

LOGO ON PLAYER'S SHORTS
"BRIDGE IMAGE"

transform the aging former base into a mix of affordable and market-rate homes designed to be sustainable, saving water and energy. The massive weight from the development will also compact the soil to keep the island stable during earthquakes, while a new seawall will protect it from sea level rise and possible tsunamis. New ramps will run to and from the new East Span and refitted West.

While the original Bay Bridge may have been criticized for its ungraceful style, it did have a place in popular culture, appearing in such movies as *The Graduate*, *Basic Instinct*, and *The Towering Inferno*. Even video games like *Resistance 2* and *Grand Theft Auto: San Andreas* included it. William Gibson used it in his futuristic *Bridge Trilogy* and it has a role in the graphic novel *How Loathsome*. The new Bay Bridge and its East Span will no doubt be equally iconic—the Golden State Warriors, the NBA team based in Oakland, already uses it as their logo *(Figures 112 & 113)*. The bridge represents the history of the Bay Area, past and present, summarizing its expansion and symbolizing its imminent growth. In its new form, it links not only Oakland and San Francisco but also future worlds to come.

THE ORIGINAL BAY BRIDGE

Official opening	November 12, 1936
Total project length (bridge and approaches)	8.4 miles
Total cost to build bridge and Transbay Terminal	$77 million
Height of West Span towers	519 feet
Size of Yerba Buena Island tunnel, largest bore tunnel in the world	76 feet wide, 58 feet high
Depth of deepest pier on the original East Span	242 feet (71 in water, 171 in mud)
Amount of cable wire used	18,500 tons
Amount of concrete	1 million cubic yards
Amount of paint	200,000 gallons
Daily traffic in 1936	21,000 vehicles, increasing to 53,000 in 1958 when the train tracks were removed
Total number of cars using the bridge from opening through June 30, 1937	5,837,835
Total number of interurban passengers from January 1, 1939 through June 30, 1939	9,639,724

THE NEW BAY BRIDGE

Length of the world's longest self-anchored suspension bridge	12,047 feet
Length of suspension cable that wraps around the roadway	nearly 1 mile
Number of individual pieces making up the Skyway	452
Tower construction began	2009
High of tower when completed	525 feet
Location of tower	607 feet from Yerba Buena Island and 1,263 feet from the beginning of the Oakland Skyway
Weight of the double-cable saddle atop the tower	ca. 450 tons

REFERENCES

BAYBRIDGEINFO.ORG: sponsored by Caltrans, the official Bay Bridge website contains up-to-the-minute developments and past milestones in the bridge construction.

BIOMITIGATION.ORG: details on the bio-mitigation process employed during the construction of the East Span.

Demoro, Harre W. *The Key Route, Part One: Transbay Commuting by Train and Ferry.* Glendale, CA: Interurban Press, 1985.

Poletti, Theresa. *Art Deco San Francisco: The Architecture of Timothy Pflueger.* Princeton: Princeton Architectural Press, 2008.

Rice, Walter and Emiliano Echeverria. *The Key System: San Francisco and the Eastshore Empire.* Mt. Pleasant, SC: Arcadia, 2007.

Sapper, Vernon. *Key System Streetcars, Transit, Real Estate and the Growth of the East Bay.* Berkeley, CA: Signature Press, 2007.

Starr, Kevin. *Endangered Dreams: The Great Depression in California.* New York: Oxford University Press, 1996.

ACKNOWLEDGMENTS

Special thanks to Noah Rosen for his research and assistance in preparing this book. Thanks as well to Clive Endress of the MTC for his participation and follow-through on the new Bay Bridge.

INDEX